Tenth Edition

INTRODUCTION TO Electric Circuits

Herbert W. Jackson
Dale Temple
Brian Kelly
Karen Craigs
Lauren Fuentes

LAB MANUAL

Karen Craigs • Lauren Fuentes

OXFORD
UNIVERSITY PRESS

Oxford University Press is a department of the University of Oxford.
It furthers the University's objective of excellence in research, scholarship,
and education by publishing worldwide. Oxford is a registered trade mark of
Oxford University Press in the UK and in certain other countries.

Published in Canada by
Oxford University Press
8 Sampson Mews, Suite 204,
Don Mills, Ontario M3C 0H5 Canada

www.oupcanada.com

Library and Archives Canada Cataloguing in Publication
Title: Introduction to electric circuits. Lab manual / Herbert W. Jackson, Dale Temple, Brian Kelly, Karen Craigs, Lauren Fuentes.
Names: Jackson, Herbert W., author. | Temple, Dale, author. | Kelly, Brian, 1948- author. | Craigs, Karen, author. | Fuentes, Lauren, author.
Description: Tenth edition. | Accompanies Introduction to electric circuits, 10th edition.
Identifiers: Canadiana 20190061588 | ISBN 9780199031467 (spiral bound)
Subjects: LCSH: Electric circuits—Laboratory manuals. | LCGFT: Laboratory manuals.
Classification: LCC TK454 .J28 2019 Suppl. | DDC 621.319/2—dc23

Cover image: TEK IMAGE/SCIENCE PHOTO LIBRARY/Getty Images
Cover design: Laurie McGregor

Oxford University Press is committed to our environment.
Wherever possible, our books are printed on paper which comes from
responsible sources.

Printed and bound in Canada

PREFACE

This manual contains a collection of experiments to accompany the text *Introduction to Electric Circuits,* Tenth Edition. It is intended to support two single-semester courses in Basic Electricity and contains thirteen DC and fifteen AC experiments. There are enough labs to cover a typical semester at a technical college. Allowing for exams and startup time, there are usually thirteen effective teaching weeks.

The experiments in this lab manual have been chosen to cover the main topics taught in foundation-level courses in electrical theory and can be done with inexpensive test equipment and circuit components. These experiments have been developed and refined over many years and are written in an easy-to-follow, step-by-step manner to accommodate students who may not be familiar with an electrical laboratory environment. There is a brief discussion at the beginning of each lab covering the theory behind the experiment to be carried out. Key equations are developed and highlighted, and a numerical example is solved.

The labs are designed to be completed in a two-hour lab slot, which usually includes enough extra time to allow students to complete their reports, if the instructor desires. With that in mind, the manual is formatted to allow for questions to be answered on the lab sheet itself, if a formal report is not required. The questions are intended to test the students' comprehension of the theoretical concepts verified by the experimental results. Each lab contains a numerical question to further test the understanding of the electrical theory being studied in the experiment.

The following page contains a complete list of the equipment and components required to perform every lab in the book.

Karen Craigs & Lauren Fuentes

EQUIPMENT AND COMPONENTS LIST

Name	Quantity	Specifications
Power Supplies		
[1]DC Power Supply	2	0-24 V(min), 3-A, regulated
AC Power Supply (Variac)	1	0-140-V, 60-Hz, 5 A
Function Generator	1	4 digit, 2 MHz, 20-V P-P
Isolation Transformer	1	115/115-V, 100-VA
Metering		
Digital Multimeter	2	Bench, 3½ digits
Digital Oscilloscope	1	20-MHz, two or more channels

[1]A single dual output power supply can be used.

Components				
[2]Resistors ½ W, 5%	47 Ω	820 Ω	2.7 kΩ	8.2 kΩ
	100 Ω	1 kΩ (2)	3.3 kΩ	10 kΩ
	220 Ω	1.2 kΩ	3.9 kΩ	15 kΩ
	330 Ω	1.5 kΩ	4.7 kΩ	100 kΩ (2)
	470 Ω	1.8 kΩ	5.6 kΩ	
	560 Ω	2.2 kΩ	6.8 kΩ	
Potentiometers, 1 W	10 kΩ	5 kΩ		
Decade Resistance Box (Optional)	5 Decades, 1 Ω steps, 1%			
Inductors (RF 50 mA)	25 mH	50 mH	100 mH	
Capacitors (Film 100 V)	0.001 μF	0.047 μF	0.22 μF	1.0 μF
	0.0047 μF	0.1 μF	0.47 μF	
Capacitors (Electrolytic 50 V)	100 μF			
[3]Meter Movement	1 mA, 100 Ω			
Breadboard/ProtoBoard	Solderless Breadboard with Jumper Wire Kit			
Line Cord	AC Plug with alligator clips			
Test Leads	5 Sets, Black + Red, Banana Plug to Alligator Clips			
	BNC to Alligator Clips			
Scope Probe	20-MHz With GND Reference Switch			
Switches	Single Pole Single Throw, 2-A, Toggle or Slide			
	Double Pole Single Throw, 2-A, Toggle or Slide			

[2] Extra resistors will be required for Experiment #1.
[3] If a 100-Ω movement is not available, a resistor can be connected in series with the movement to increase its resistance to 100 Ω.

CONTENTS

EXPERIMENT

1

RESISTOR COLOUR CODE AND THE OHMMETER

OBJECTIVES

1. To become familiar with the resistor colour code and use it for determining the value of resistors.
2. To use an ohmmeter for measuring resistor values.
3. To become familiar with the potentiometer.

DISCUSSION

Resistor Colour Code

Carbon resistors have colour bands used to encode the resistor value. Colour bands are used because they are more easily read after long service, especially if the resistor overheats.

Each colour corresponds to a number as per the following table.

Black	Brown	Red	Orange	Yellow	Green	Blue	Violet	Grey	White	Silver	Gold
0	1	2	3	4	5	6	7	8	9	0.01	0.1

multiplier only

The colours are converted into numbers and interpreted as shown in the figures below, starting from the side closest to the first colour band. Five-band resistors typically follow the figure on the left, for precision, and four-band resistors typically follow the figure on the right, eliminating the band for reliability. The tolerance band is often physically set apart from the others, and is sometimes a little thicker.

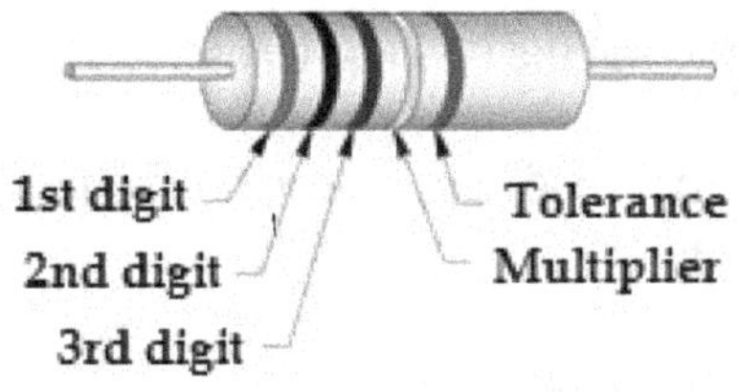

1st digit
2nd digit
3rd digit
Tolerance
Multiplier

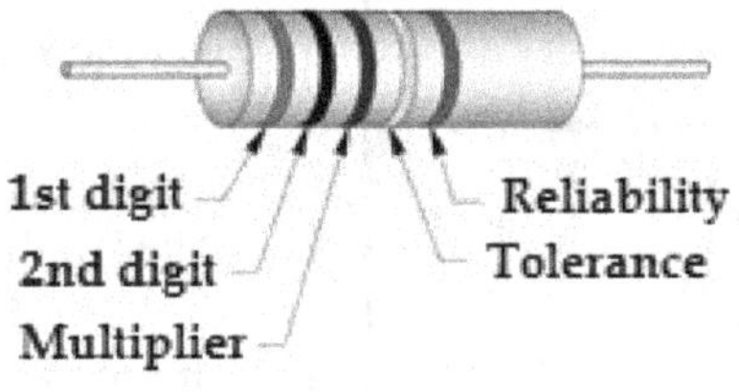

1st digit
2nd digit
Multiplier
Reliability
Tolerance

Tolerance: Maximum % deviation from the colour coded value. The following colours are used:

No colour	Silver	Gold	Red	Brown
20%	10%	5%	2%	1%

Green	Blue	Violet	Grey
0.50%	0.25%	0.10%	0.05%

Reliability: Percentage of failure per 1000 hours of use. The following colours are used:

Brown	Red	Orange	Yellow
1%	0.1%	0.01%	0.001%

Example: What information can be determined from a reliability resistor with the colour code brown-black-red-gold-brown?

$\text{Value} = 10 \times 10^2 = \mathbf{1000\Omega}$, Tolerance = 5%, which means the actual resistance value will lie within ±5% of the coded, i.e. between 950 Ω and 1050 Ω.

Reliability = 1%, which means 1 out of every 100 resistors will not lie within the tolerance range after 1000 hours of operation at its rated power.

Analog Ohmmeter

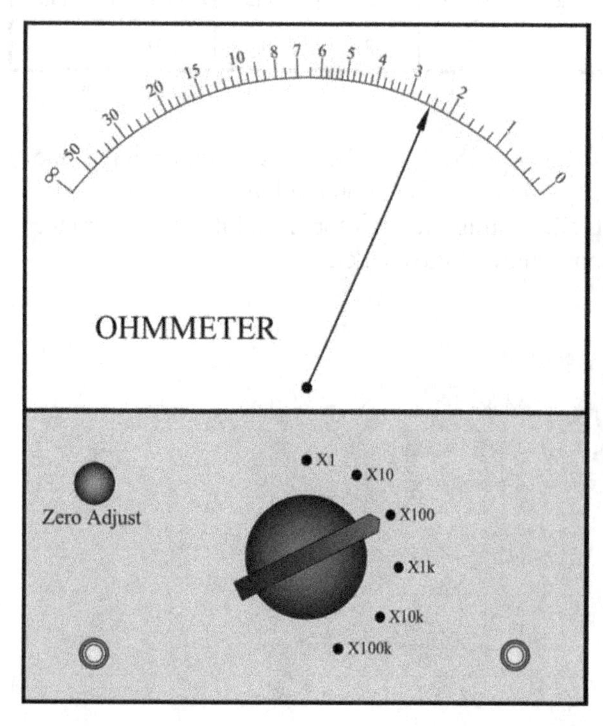

Using the Analog Ohmmeter

Select the appropriate multiplier range for resistance being measured. The pointer should lie between one quarter and three quarter full-scale deflection for accurate measurement.

Zero the ohmmeter by short circuiting the terminals with a test lead and adjusting the zero control until the pointer indicates 0 Ω.

Connect the resistor between the two test leads.

Read the value off the scale and apply the appropriate multiplier to the reading.

The ohmmeter has its own internal power source (batteries) and should never be connected to a circuit that already has power.

Digital Multimeter (Ω Function)

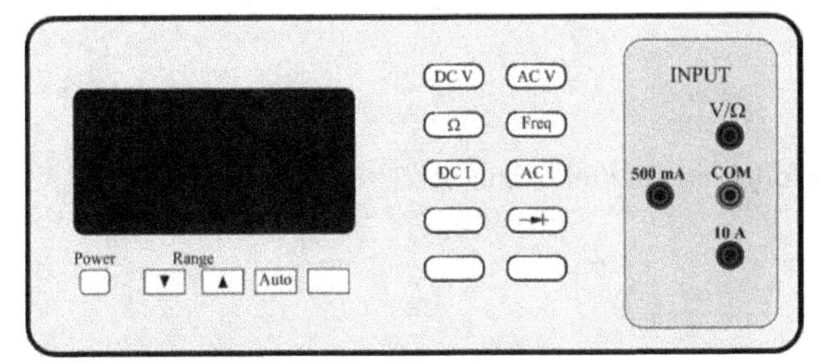

Press the **Ω** function button. Press the autorange button if available. If not, select the appropriate range for the resistance being measured. If the range is too high, the measurement may be rounded off and accuracy will be lost.

Connect test leads to the V/Ω and COM terminals. Make sure there is no power on the circuit when resistance measurements are being made.

Variable Resistors

Carbon-film or wire-wound.

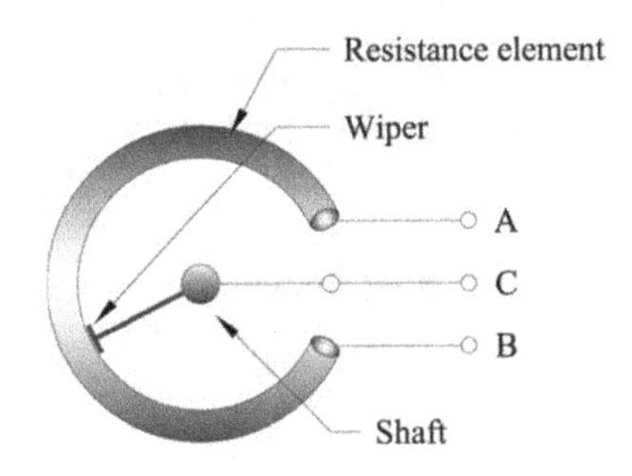

The movable wiper arm makes contact with a portion of the resistance.

Potentiometer – 3-terminal variable resistor

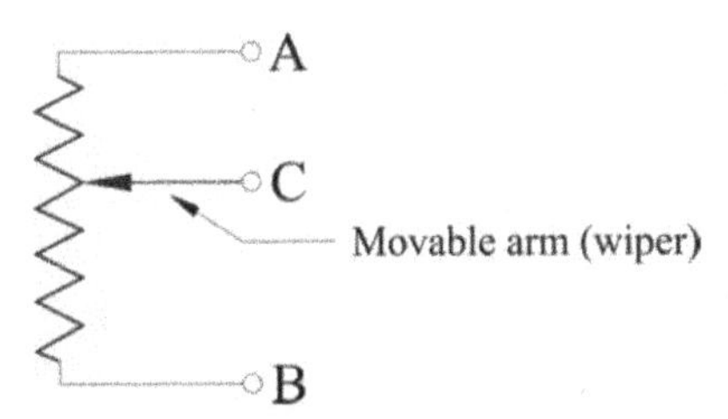

Resistance between A & B is constant. Resistance between A & C and C & B varies with position of wiper.

$R_{AC} + R_{CB} = R_{AB}$ = rated potentiometer resistance.

Rheostat – 2-terminal variable resistor

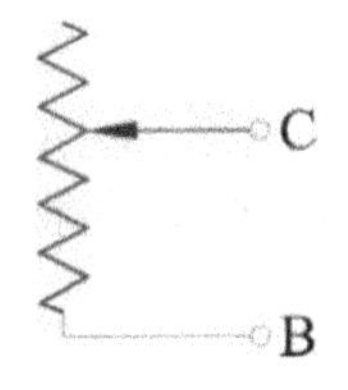

Resistance R_{BC} varies from 0 Ω to rated resistance of rheostat.

EQUIPMENT

Instruments:

- Digital Multimeter set on Ohms function

Resistors: (5%, ½ W or higher)

- 10 assorted 4- and 5-band resistors with the following multiplier bands:
  - 1 with Black
  - 2 with Brown
  - 3 with Red
  - 3 with Orange
  - 1 with Yellow
- 10-kΩ potentiometer

PROCEDURE

1. Using the resistor colour code, determine the value and percentage tolerance of each of the 10 resistors you have chosen and record in Table 1-1.
2. Select a resistor for measurement, and using the coded value as a guide, select an appropriate multiplier scale on the ohmmeter. Measure the value of this resistor and record in Table 1-1.
3. Repeat step 2 for the other resistors.
4. Calculate the percentage error for each of the resistors in Table 1-1, using the following formula:

$$\%\text{Error} = \left[\frac{|\text{Coded Value} - \text{Measured Value}|}{\text{Coded Value}}\right] \times 100\%$$

5. Examine the 10-kΩ potentiometer and orient it as shown in Figure 1-1.

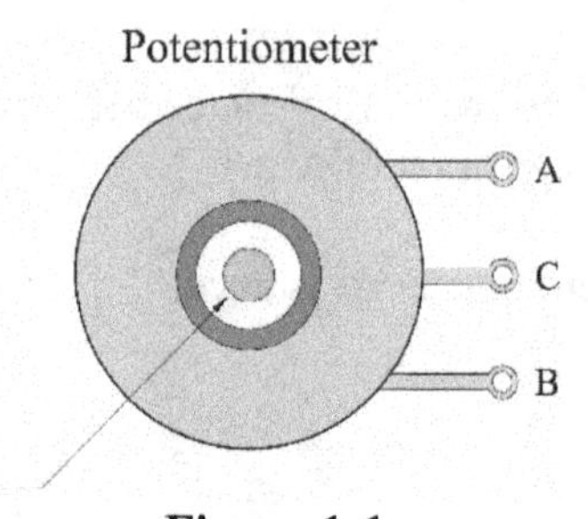

Figure 1-1

6. Rotate the shaft to a fully clockwise position.
7. Using your ohmmeter on an appropriate scale, measure the resistance between terminals A & B (R_{AB}) Record this in Table 1-2. Repeat this between terminals A & C (R_{AC}) and terminals B & C (R_{CB}). Record your results in Table 1-2.
8. Rotate the shaft to a fully counterclockwise position and repeat step 7.
9. Rotate the shaft to approximately centre position and repeat step 7.

RESULTS

Table 1-1 Resistor Colour Code

	RESISTORS									
	1	2	3	4	5	6	7	8	9	10
1st Colour Band										
2nd Colour Band										
3rd Colour Band										
4th Colour Band										
5th Colour Band										
Coded Value Ω										
Tolerance %										
Meas. Value Ω										
%Error										

Table 1-2 Potentiometer Resistance Measurements

Potentiometer Shaft Position	MEASURED VALUES			CALCULATED VALUE
	R_{AB} (Ω)	R_{AC} (Ω)	R_{CB} (Ω)	$R_{AB} = R_{AC} + R_{CB}$ (Ω)
Fully CCW				
Fully CW				
Centre				

QUESTIONS

1. Complete the following table:

Resistor Value	First Colour Band	Second Colour Band	Third Colour Band	Fourth Colour Band
	brown	red	red	silver
	orange	orange	brown	gold
	yellow	violet	orange	silver
	red	red	black	no colour
	grey	red	silver	gold
560 Ω, 2%				
10 kΩ, 5%				
0.11 Ω, 1%				
29 Ω, 10%				
7100 Ω, 20%				

2. An ohmmeter is set on the *R* x 10 scale and reads 1500 Ω. The scale is then changed to the *R* x 1 scale.

(a) Describe the procedure you need to follow to ensure the meter indicates an accurate reading.

Digital Meter: Describe what must be done to ensure maximum accuracy is achieved.

(b) Would the new reading be more or less accurate than the original reading? Why?

____

____

____

____

____

____

3. Describe the effect of varying the movable arm of the potentiometer on the resistance between terminals A & B (R_{AB}). Explain why this happens.

____

____

____

____

____

____

____

____

EXPERIMENT

2

OHM'S LAW

OBJECTIVES

1. To verify the relationship between voltage, current, and resistance in an electric circuit.
2. To become familiar with the digital voltmeter and ammeter.

DISCUSSION

Ohm's Law

This is the basic relationship between ***E***, ***I***, and ***R*** in an electrical circuit.

In any circuit, the current flowing is directly proportional to the EMF and inversely proportional to the resistance.

$$Current = \frac{Voltage}{Resistance} \quad \text{or} \quad I = \frac{E}{R}$$

Example 1: If a series circuit has a battery with an EMF of 12 V and a lamp with resistance 24 Ω, find the current flowing in the circuit.

$$I = \frac{E}{R} = \frac{12\,\text{V}}{24\,\Omega} = \mathbf{0.5\,A}$$

Example 2: How much voltage is necessary to send a current of 10 mA through a 5-kΩ resistance?

$$I = \frac{E}{R} \;\Rightarrow\; E = I \times R = \left(10 \times 10^{-3}\,\text{A}\right) \times \left(5 \times 10^{3}\,\Omega\right) = \mathbf{50\,V}$$

Example 3: What is the total resistance present in a circuit when a source of 100 V produces a current of 50 μA?

$$I = \frac{E}{R} \;\Rightarrow\; R = \frac{E}{I} = \frac{100\,\text{V}}{50 \times 10^{-6}\,\text{A}} = 2 \times 10^{6}\,\Omega = \mathbf{2\,M\Omega}$$

Digital Multimeter (Voltage Function)

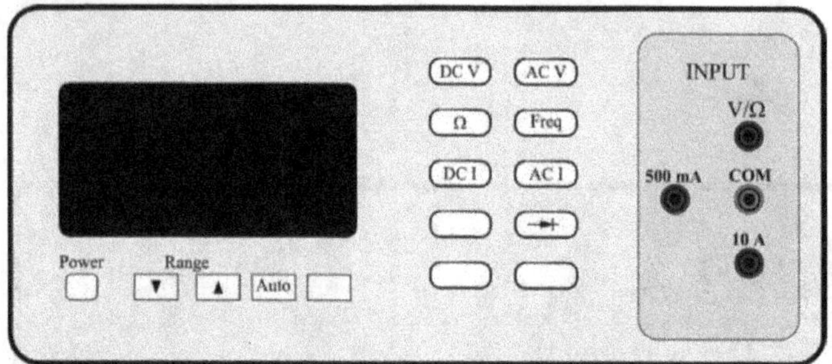

Press the **DC V** function button. Press the autorange button, if available. If not, select the appropriate range for the component being measured. If the range is too high, the measurement may be rounded off and accuracy will be lost.

Connect test leads to the V/Ω and COM terminals.

The voltmeter is always connected in **parallel** with the component having its voltage measured. In the image shown below, the voltmeter is connected across resistor R_2 and will display its corresponding voltage drop.

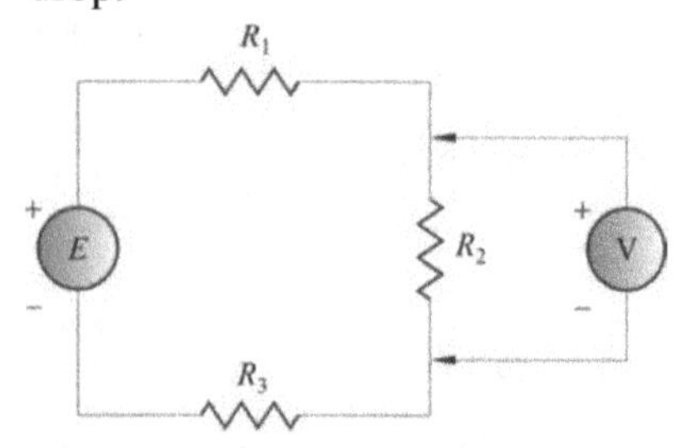

Digital Multimeter (Ammeter Function)

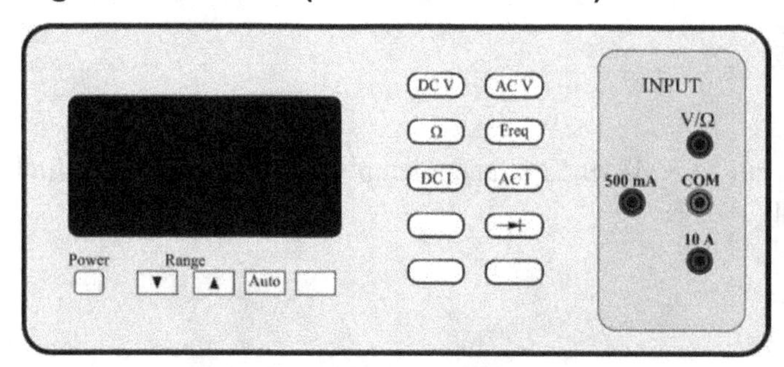

Press the **DC I** function button. Press the autorange button, if available. If not, select the appropriate range for the component being measured. If the range is too high, the measurement may be rounded off and accuracy will be lost.

Connect test leads to the 500 mA and COM terminals.

The ammeter is always connected in **series** with the component having its current measured. In the image shown below, the ammeter is connected in series before resistor R_2 and will display the circuit's current.

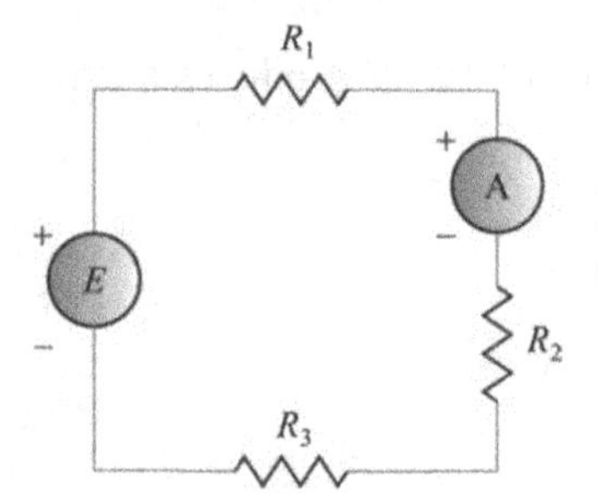

Ensure the battery output is **off** before inserting the ammeter in series to avoid making accidental pathways in the circuit.

EQUIPMENT

Power Supplies:
- Variable 0-24 V DC, regulated

Instruments:
- Two Digital Multimeters (DMM)

Resistors: (5%, ½ W or higher)
- One each of 1-kΩ, 2.2-kΩ, 3.3-kΩ, and 4.7-kΩ

PROCEDURE

1. Using the 1-kΩ resistor for ***R***, connect the circuit shown in Figure 2-1. Set up one DMM as a voltmeter and the other as an ammeter, and insert them into the positions shown. Set up the power supply to 2 volts (with a current limit of 15 mA if using a digital supply), and keep the output **off** for now. **Have an instructor check the circuit before applying power.**

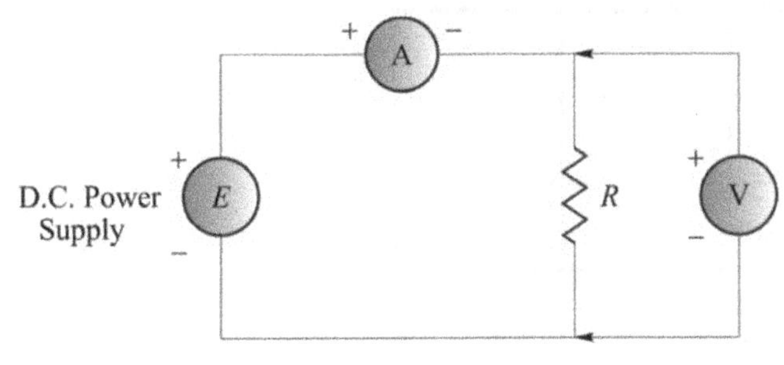

Figure 2-1

2. Turn **on** the power supply. Measure the current with the ammeter and record the value in Table 2-1.

3. Adjust the power supply for the voltages shown in Table 2-1 and record the corresponding currents.

4. Turn **off** the power. Replace the 1-kΩ resistor with the 2.2-kΩ resistor.

5. Adjust the power supply to the voltages shown in Table 2-2 and record the corresponding currents.

6. Repeat step 4 for resistances of 3300 Ω and 4700 Ω. Adjust voltages to the values shown in Tables 2-3 and 2-4. Record corresponding currents in each case.

7. Calculate the theoretical value of currents from Ohm's law and record your answers in the tables.

RESULTS

Table 2-1 Ohm's Law

R	**1000 Ω**				
V **(volts)**	2	4	6	8	10
I **(milliamps)**					
I **(calculated)**					

Table 2-2

R	**2200 Ω**				
V **(volts)**	3	6	9	12	15
I **(milliamps)**					
I **(calculated)**					

Table 2-3

R	**3300 Ω**				
V **(volts)**	4	8	12	16	20
I **(milliamps)**					
I **(calculated)**					

Table 2-4

R	**4700 Ω**				
V **(volts)**	5	10	15	20	24
I **(milliamps)**					
I **(calculated)**					

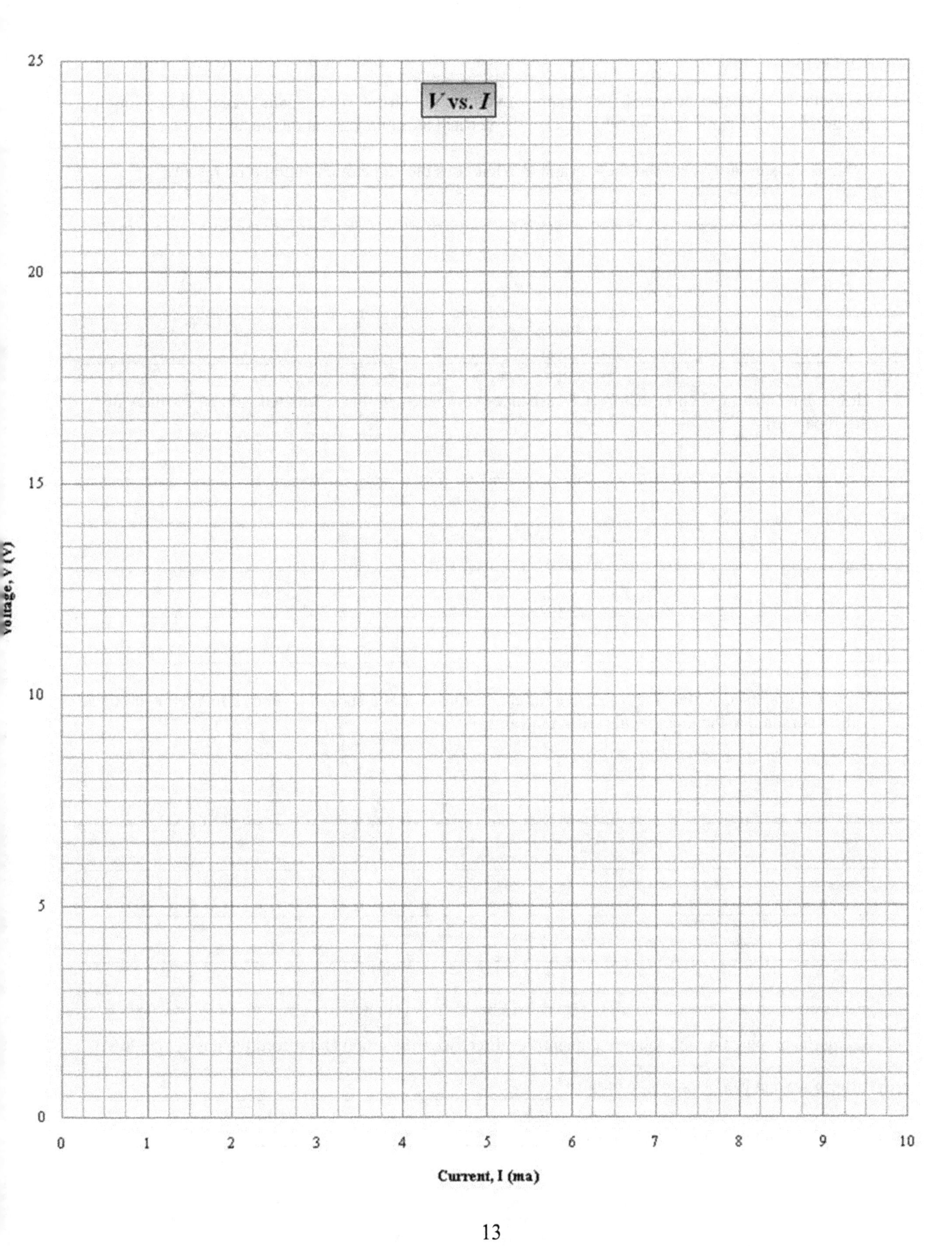

V vs. I
Voltage, V (V)
Current, I (ma)
0
5
10
15
20
25
0
1
2
3
4
5
6
7
8
9
10

QUESTIONS

1. Using the graph paper provided, plot graphs of voltage vs. current for the data in each table. Plot all the graphs on the same axes with voltage on the vertical axis and current on the horizontal axis.

What is the similarity between these graphs? What does the slope of each graph represent?

____

____

____

____

2. How do the measured currents for each resistance compare with the calculated values? Explain any discrepancies.

____

____

____

____

____

3. From the graphs, calculate the resistance for each value of R at a current level of 5 mA. How do these values compare to the theoretical resistance values?

____

____

____

____

____

____

4. Using the graphs, find the following quantities directly on the graph and record the answers below.

(a) With $R = 1000\ \Omega$ and $I = 5$ mA, find V: ______________________________

(b) With $R = 3300\ \Omega$ and $V = 20$ V, find I: ______________________________

EXPERIMENT

3

RESISTANCE IN SERIES

OBJECTIVES

1. To determine the rule for the total resistance of series-connected resistances.
2. To verify total resistance of a series circuit by Ohm's law.

DISCUSSION

When circuit components are connected so that there is only one path for current to flow, they are said to be in *series*.

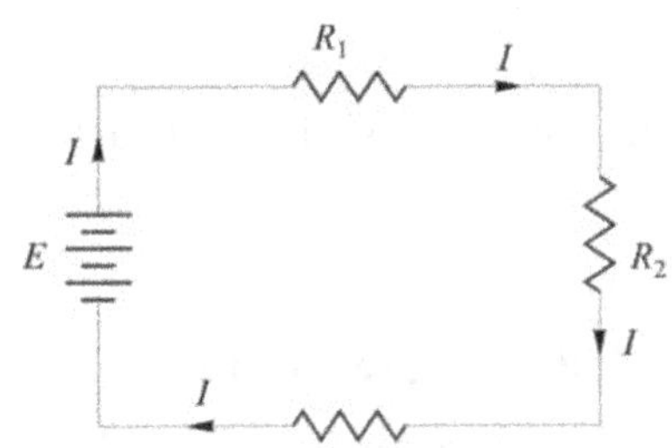

Rules for the series circuit

1. Current is the same through each component.
2. $R_T = R_1 + R_2 + R_3$
3. $I = \dfrac{E}{R_T}$

Example: If in the circuit shown, $E = 24$ V, $R_1 = 2\ \Omega$, $R_2 = 4\ \Omega$ and $R_3 = 6\ \Omega$, calculate the total current and voltage across each resistor.

$$R_T = 2\Omega + 4\Omega + 6\Omega$$

$$I = \frac{E}{R_T} = \frac{24\text{ V}}{12\ \Omega} = \mathbf{2\ A}$$

Voltage drops may be found by applying Ohm's law to each individual resistor.

$$V = I\,R$$
$$V_{2\Omega} = 2\text{ A} \times 2\ \Omega = 4\text{ V}$$
$$V_{4\Omega} = 2\text{ A} \times 4\ \Omega = 8\text{ V}$$
$$V_{6\Omega} = 2\text{ A} \times 6\ \Omega = 12\text{ V}$$

Note when the voltage drops are added up, the total is 24 V. This proves **Kirchhoff's Voltage Law,** which states:

The sum of the voltage drops in a series circuit equals the total applied voltage.

$$V_1 + V_2 + V_3 + \cdots\cdots\cdots = V_T$$

EQUIPMENT

Power Supplies:
- Variable 0-24 V DC, regulated

Instruments:
- Digital Multimeter (DMM)

Resistors: (5%, ½ W or higher)
- One each of 330-Ω, 560-Ω, 1500-Ω, 2700-Ω, 3900-Ω and 5600-Ω

PROCEDURE

PART A: Total Resistance of Series Connected Resistors (Ohmmeter Method)

1. Measure the actual resistance of your resistors with your DMM and record the results in Table 3-1.

2. Connect R_1 and R_2 in series as shown in Figure 3-1 below. Measure the total resistance across the combination and record in Table 3-2.

330 Ω 560 Ω

Figure 3-1

3. Repeat step 2 for the other series combinations shown in Table 3-2.

4. Using the *measured* values of resistance from Table 3-1, calculate the total resistance of each combination and enter the values in Table 3-2.

PART B: Total Resistance of Series Connected Resistors (Ohm's Law Method)

5. Connect the circuit shown in Figure 3-2 using combination 1 from Table 3-2.

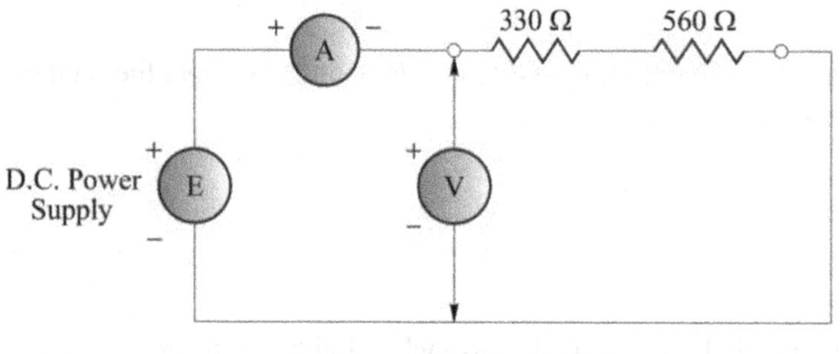

Figure 3-2

6. Turn **on** the power supply and adjust the voltage to 15 V (with a current limit of 20 mA if using a digital supply). Turn **on** the power supply output and measure the current in the circuit. Convert the current into amperes and record the value in Table 3-3.

7. Repeat step 6 with the other series combinations from Table 3-2 and record the results in Table 3-3.

8. Calculate the total resistance from the *measured* values of voltage and current and record the results in Table 3-3.

RESULTS

Table 3-1 Measured Resistances

Resistor→	R_1	R_2	R_3	R_4	R_5	R_6
Coded Value	330 Ω	560 Ω	1.5 kΩ	2.7 kΩ	3.9 kΩ	5.6 kΩ
Measured Value						

Table 3-2 Total Resistance of Series Circuits (Ohmmeter Method)

Comb	**Resistors in Series**					**Total Resistance (Ohmmeter) (Ω)**	**Total Resistance (Formula) (Ω)**
1	330 Ω	560 Ω					
2	1.5 kΩ	2.7 kΩ					
3	3.9 kΩ	5.6 kΩ					
4	330 Ω	560 Ω	1.5 kΩ				
5	1.5 kΩ	2.7 kΩ	3.9 kΩ				
6	330 Ω	560 Ω	1.5 kΩ	2.7 kΩ			
7	330 Ω	560 Ω	1.5 kΩ	2.7 kΩ	3.9 kΩ		

Table 3-3 Total Resistance of Series Circuits (Ohm's Law Method)

Comb	**Applied Voltage (V)**	**Measured Current (A)**	R_T **Ohm's Law** $R_T = E/I$ **(Ω)**	R_T **Ohmmeter (from Table 3-2) (Ω)**	R_T **Formula (from Table 3-2) (Ω)**
1	15				
2	15				
3	15				
4	15				
5	15				
6	15				
7	15				

QUESTIONS

1. Describe two methods used in this experiment to find, by measurement, the total resistance of series-connected resistors. Which method would you consider to be more accurate? Explain your answer.

2. How do the measured and calculated values of total resistance compare in Table 3-2? Explain any discrepancies.

3. Why was the resistance of each individual resistor measured?

____

____

____

____

4. If one of the resistors in a series circuit burns out and becomes an open circuit, what happens to the current in the circuit? Explain your answer.

____

____

____

____

____

____

5. A set of Christmas lights consists of 20 5-Ω lamps connected in series to a 120-V supply. How much current will flow and how much voltage will there be across each lamp?

____

____

____

____

____

____

____

____

____

EXPERIMENT

4

RESISTANCE IN PARALLEL

OBJECTIVES

1. To determine the rule for the total resistance of parallel-connected resistances.
2. To verify the total resistance of a parallel circuit by Ohm's law.

DISCUSSION

When circuit components are connected so that the current has more than one possible path, they are said to be in *parallel.*

Rules for the Parallel Circuit

1. Since points A, B, and C are all at the same potential, therefore the voltage across each component in a parallel branch is the same. ⇨ $V_1 = V_2 = V_3 = E$
2. Total resistance is found using the **reciprocal formula**. ⇨ $\frac{1}{R_T} = \frac{1}{R_1} + \frac{1}{R_2} + \frac{1}{R_3} + \cdots\cdots$
3. $I_T = \frac{E}{R_T}$

Example: Find the total resistance and current in the previous circuit if $E = 24$ V, $R_1 = 2\ \Omega$, $R_2 = 4\ \Omega$, and $R_3 = 6\ \Omega$.

$$\frac{1}{R_T} = \frac{1}{2\Omega} + \frac{1}{4\Omega} + \frac{1}{6\Omega}$$

$$= 0.5 + 0.25 + 0.167 = 0.917\text{S}$$

$$R_T = \frac{1}{0.917\text{S}} = \mathbf{1.09\ \Omega}$$

$$I_T = \frac{E}{R_T} = \frac{24V}{1.09\Omega} = \mathbf{22\ A}$$

Note that the total resistance is **always smaller** than the smallest resistor in the parallel combination.

Ohm's law can now be used to find the individual branch currents.

$$I_1 = \frac{E}{R_1} = \frac{24\text{ V}}{2\Omega} = 12\text{ A}$$

$$I_2 = \frac{E}{R_2} = \frac{24\text{ V}}{2\Omega} = 6\text{ A}$$

$$I_3 = \frac{E}{R_3} = \frac{24\text{ V}}{6\Omega} = 4\text{ A}$$

Note when the branch currents are added up the total is 22 A. This proves **Kirchhoff's Current Law,** which states:

The sum of the branch currents in a parallel circuit equals the total current entering the branches.

$$I_1 + I_2 + I_3 + \cdots\cdots = I_T$$

EQUIPMENT

Power Supplies:
- Variable 0-24 V DC, regulated

Instruments:
- Digital Multimeter (DMM)

Resistors: (5%, ½ W or higher)
- One each of 820-Ω, 1-kΩ, 2.2-kΩ, 3.3-kΩ, and 4.7-kΩ

PROCEDURE

PART A: Total Resistance of Parallel Connected Resistors (Ohmmeter Method)

1. Measure the actual resistances of your resistors using the ohmmeter function on the DMM and record the results in Table 4-1.

2. Connect R_1 and R_2 in parallel as shown in Figure 4-1. Measure the total resistance across the combination and record the result in Table 4-2.

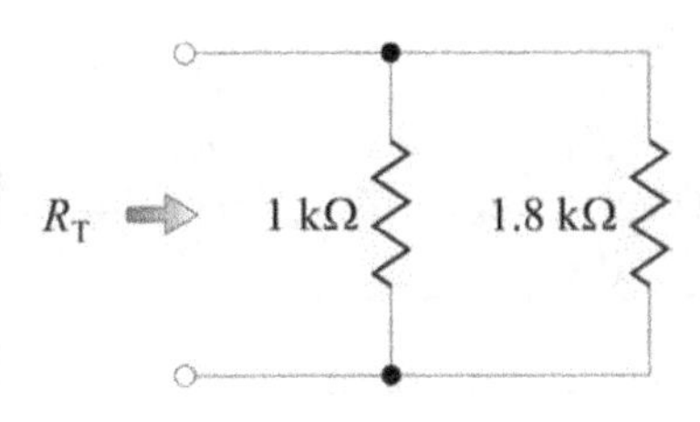

Figure 4-1

3. Repeat step 2 for the other parallel combinations shown in Table 4-2.

4. Using the *measured* values of resistance from Table 4-1, calculate the total resistance of each combination and enter the values in Table 4-2.

PART B: Total Resistance of Parallel Connected Resistors (Ohm's Law Method)

5. Connect the circuit shown below in Figure 4-2 using combination 1 from Table 4-2.

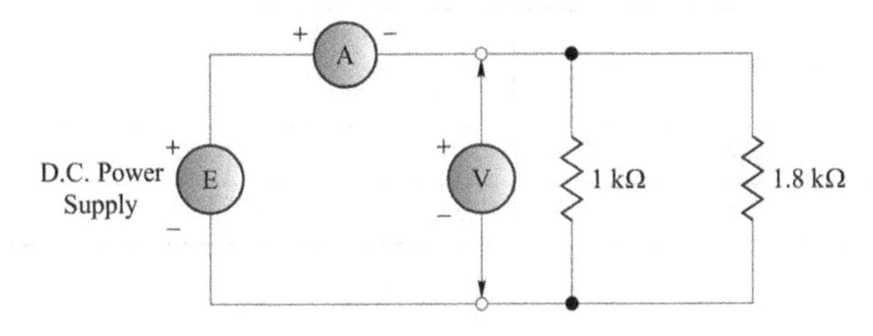

Figure 4-2

6. Turn **on** the power supply and adjust the voltage to 15 V (with a current limit of 30 mA if using a digital supply). Measure the current in the circuit. Convert the current into amps and record the value in Table 4-3.

7. Repeat step 6 with the other parallel combinations from Table 4-2 and record the results in Table 4-3.

8. Calculate the total resistance from the *measured* values of voltage and current and record in Table 4-3.

RESULTS

Table 4-1 Measured Resistances

Resistor→	R_1	R_2	R_3	R_4	R_5
Coded Value	1 kΩ	1.8 kΩ	2.7 kΩ	3.3 kΩ	3.9 kΩ
Measured Value					

Table 4-2 Total Resistance of Parallel Circuits (Ohmmeter Method)

Comb.	Resistors in Parallel					R_T (Ohmmeter) (Ω)	R_T (Formula) (Ω)
1	1 kΩ	1.8 kΩ					
2	1 kΩ	1.8 kΩ	2.7 kΩ				
3	1 kΩ	1.8 kΩ	2.7 kΩ	3.3 kΩ			
4	1 kΩ	1.8 kΩ	2.7 kΩ	3.3 kΩ	3.9 kΩ		

Table 4-3 Total Resistance of Parallel Circuits (Ohm's Law Method)

Comb.	E Applied Voltage (V)	I Measured Current (A)	R_T Ohm's Law $R_T = E/I$	R_T Ohmmeter (from Table 4-2) (Ω)	R_T Formula (from Table 4-2) (Ω)
1	15				
2	15				
3	15				
4	15				

QUESTIONS

1. Describe (in words) the reciprocal formula used to calculate the total resistance of a parallel circuit.

2. How do the measured and calculated values of total resistance compare in Table 4-2? Explain any discrepancies.

3. Describe three methods used to determine the total resistance of resistors connected in parallel.

4. Using your results, describe the effect that increasing the number of resistors in parallel has on the total resistance. Explain why this happens.

5. Suppose 100-Ω and 150-Ω resistors are connected in parallel to a 120-V source. How much resistance must be placed in parallel with this combination for the circuit to draw 3 A from the source?

EXPERIMENT

5

KIRCHHOFF'S VOLTAGE AND CURRENT LAWS

OBJECTIVES

1. To verify Kirchhoff's voltage law for a closed circuit.
2. To verify Kirchhoff's current law for parallel branches.

DISCUSSION

Kirchhoff's Voltage Law: *In any closed circuit (loop) the algebraic sum of the applied voltages and voltage drops is zero.*

Sign Conventions

Voltage source Voltage arrow points from negative to positive.

Voltage drop Assume a current direction, voltage arrow points towards the end of the component where the current enters.

Drawing arrows Each source voltage and voltage drop is given a sign using the following technique. Trace around the circuit in the direction of assumed current. If arrow tail is met, sign is positive. If arrow tip is met, sign is negative.

Example: Write KVL for the circuit shown.

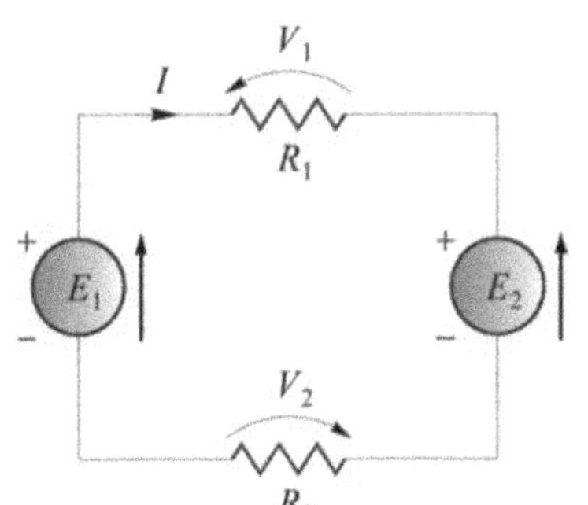

$$-V_1 - E_2 - V_2 + E_1 = 0$$

Rearranging and applying Ohm's Law

$$E_1 - E_2 - IR_1 - IR_2 = 0$$

Kirchhoff's Current Law: *The algebraic sum of the currents meeting at a junction (node) is zero.*

Sign Convention

Currents *towards* node – positive

Currents *away* from node – negative

Example: Write KCL for the following node.

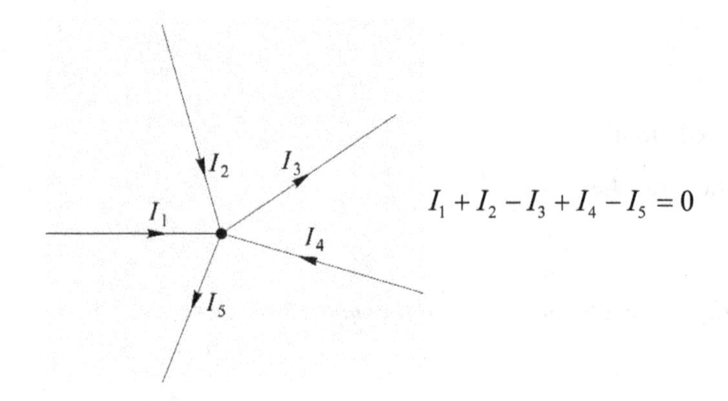

$$I_1 + I_2 - I_3 + I_4 - I_5 = 0$$

EQUIPMENT

Power Supplies:

- Variable 0–24-V DC, regulated

Instruments:

- Digital Multimeter (DMM)

Resistors: (5%, ½ W or higher)
One each of 470-Ω, 1-kΩ, 1.8-kΩ, 2.2-kΩ, 2.7-kΩ, 3.9-kΩ, 4.7-kΩ, and 5.6-kΩ

PROCEDURE

PART A: Kirchhoff's Voltage Law

1. Connect the circuit show in Figure 5-1. Turn **on** the power supply but keep the output **off**. Adjust the source voltage to 20 V (with a current limit of 15 mA if using a digital supply).

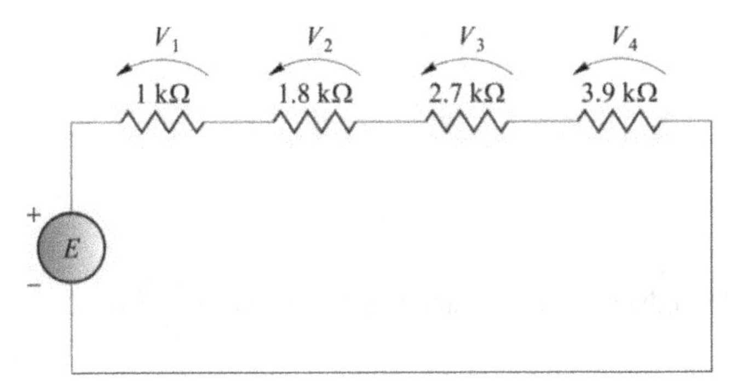

Figure 5-1

2. Measure the voltage across each resistor. Record these values in Table 5-1. Calculate the sum of these voltages and record in Table 5-1.

3. Connect the circuit shown in Figure 5-2. Use the same power supply settings as in step 1.

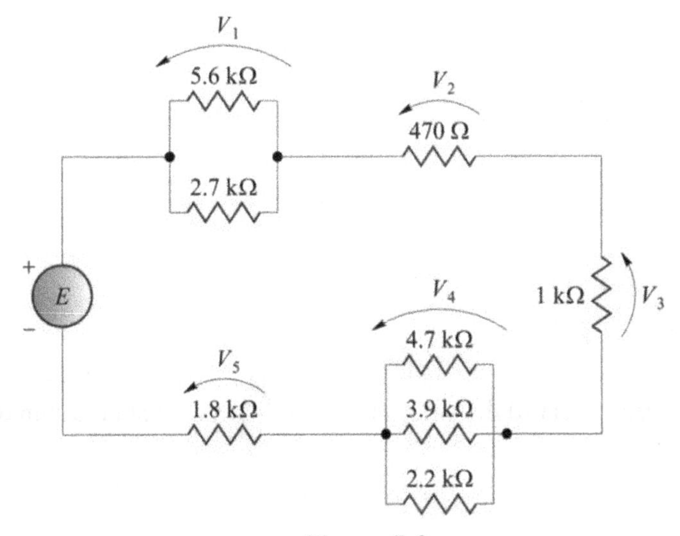

Figure 5-2

4. Measure V_1, V_2, V_3, V_4, and V_5. Record these values in Table 5-1. Calculate the sum of these voltages and record in Table 5-1.

PART B: Kirchhoff's Current Law

5. Connect the circuit shown in Figure 5-3. Adjust the power supply source voltage to 15 V.

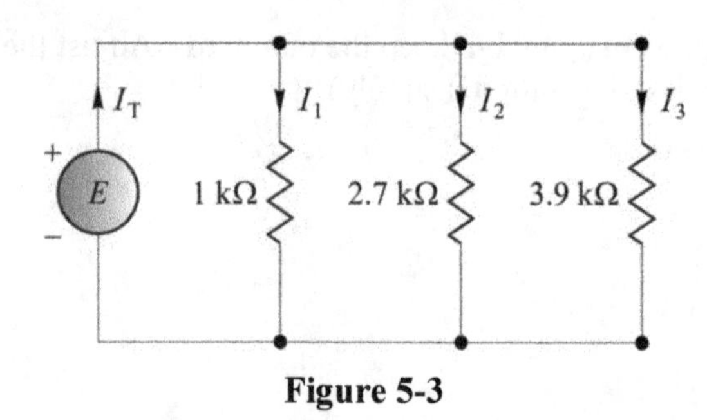

Figure 5-3

6. Measure the total current and the current through each resistor, and record the values in Table 5-2. Calculate the sum of I_1, I_2, and I_3.

7. Connect the circuit shown in Figure 5-4. Adjust the power supply source voltage to 20 V.

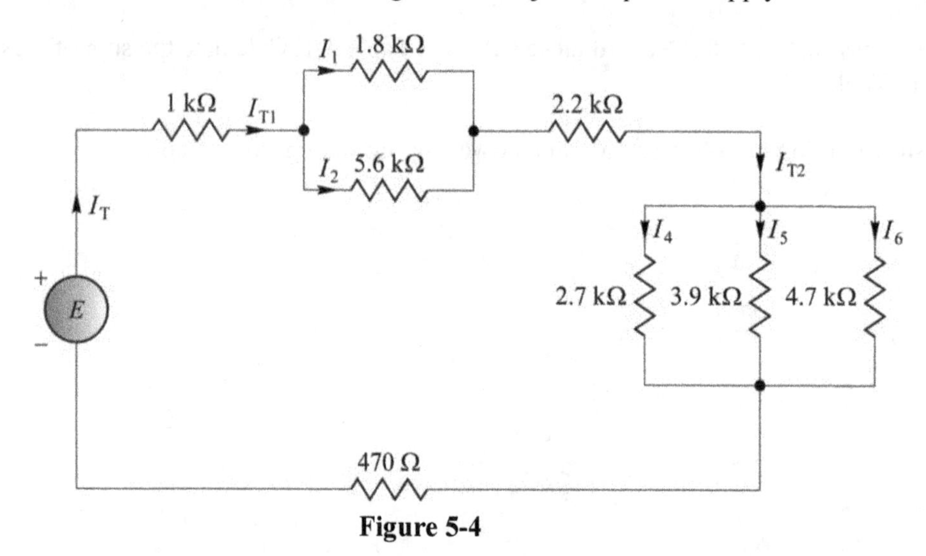

Figure 5-4

8. Measure all of the currents shown in the circuit. Record the values in Table 5-2. Calculate the sum of I_1 and I_2, and the sum of I_4, I_5, and I_6.

RESULTS

Table 5-1 Kirchhoff's Voltage Law

Circuit	E	V_1	V_2	V_3	V_4	V_5	ΣV's
Fig. 5-1							
Fig. 5-2							

Table 5-2 Kirchhoff's Current Law

Milliammeter Position	Measured Current (mA)	
	Fig. 5-3	Fig. 5-4
I_T		
I_{T1}		
I_1		
I_2		
I_3		
I_{T2}		
I_4		
I_5		
I_6		
$I_1 + I_2$		
$I_1 + I_2 + I_3$		
$I_4 + I_5 + I_6$		

QUESTIONS

1. In your own words, explain the relationship between voltage drops in a closed circuit and applied voltage. Express this relationship as a mathematical equation.

2. Is the relationship stated in Question 1 confirmed from your results? Refer to your experimental results and explain any discrepancies.

3. In your own words, explain the relationship between branch currents and total current in a parallel circuit. Express this relationship as a mathematical equation.

4. Is the relationship stated in Question 3 confirmed from your results? Refer to your experimental results and explain any discrepancies.

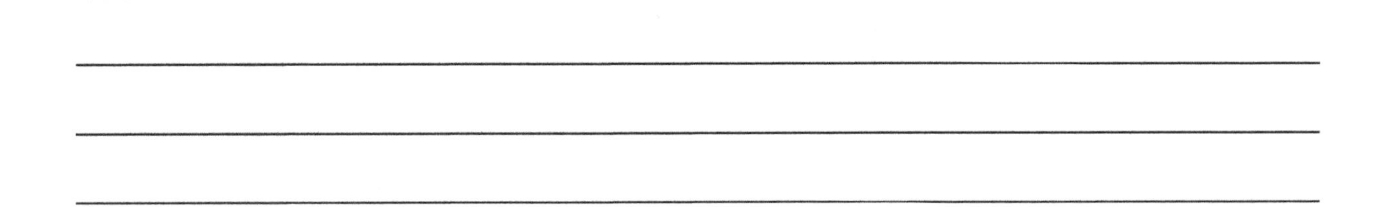

5. Using Kirchhoff's Laws, calculate the value of R_x in the circuit shown below.

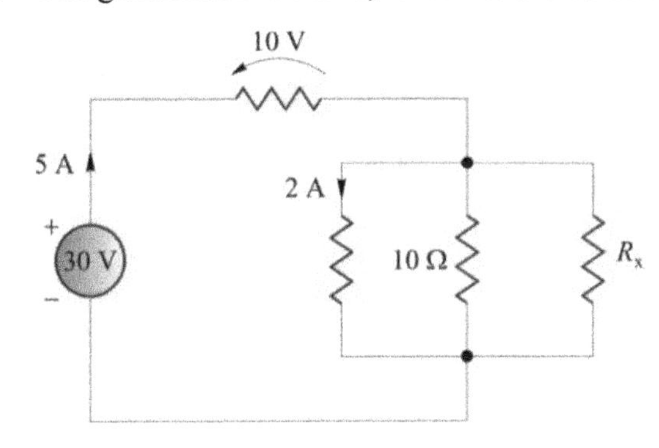

[illegible]

[illegible]

EXPERIMENT

6

THE SERIES-PARALLEL CIRCUIT

OBJECTIVES

1. To verify, by experiment, the rules for determining the total resistance of series-parallel circuits.
2. To design a series-parallel circuit meeting specified voltage and current requirements.

DISCUSSION

Series-Parallel Circuits

A circuit containing both series and parallel branches is called a *series-parallel* or *combination* circuit.

To find the total resistance of such a circuit, it is necessary to reduce the circuit systematically by combining the series and parallel sub-circuits.

Example: For the circuit shown below, calculate the total resistance.

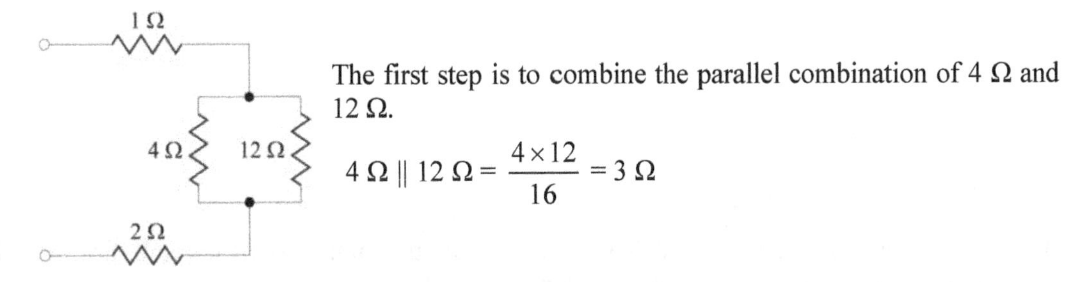

The first step is to combine the parallel combination of 4 Ω and 12 Ω.

$$4\ \Omega \parallel 12\ \Omega = \frac{4 \times 12}{16} = 3\ \Omega$$

The circuit can now be reduced to:

1 Ω

3 Ω R_{eq}

2 Ω

Applying the series circuit rule: ⇨ $R_T = \mathbf{6\Omega}$

EQUIPMENT

Power Supplies:
- Variable 0-24 V DC, regulated

Instruments:
- Digital Multimeter (DMM)

Miscellaneous:
- SPST switch

Resistors: (5%, ½ W or higher)
One each of 330-Ω, 470-Ω, 560-Ω, 1.2-kΩ, 2.2-kΩ, 3.3-kΩ, and 4.7-kΩ

PROCEDURE

PART A: Resistance of a Series-Parallel Circuit

1. Measure the actual resistance of your resistors with the ohmmeter function of your DMM and record the results in Table 6-1.

2. Connect the series-parallel circuit shown in Figure 6-1.

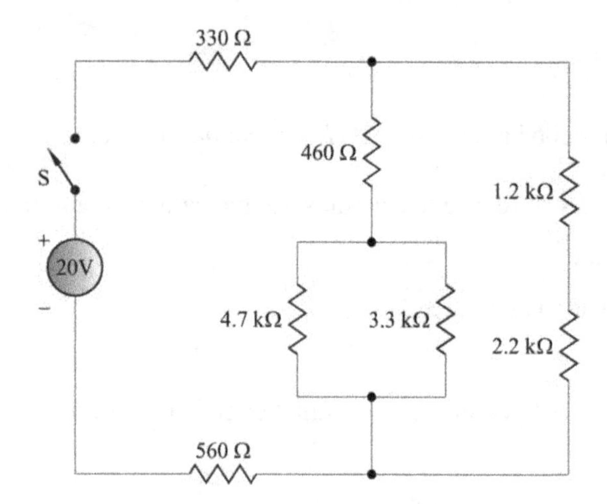

Figure 6-1

3. With power supply **off** and switch S **open**, measure the total resistance of the series-parallel circuit with your DMM. Record the total resistance in Table 6-1.

4. Using the *measured* values of resistance from Table 6-1, calculate the total resistance of the series-parallel circuit and record this in Table 6-1.

5. Turn **on** the power supply and adjust the applied voltage to 20 V (with a current limit of 15 mA if using a digital supply). Turn **on** the voltage output and **close** switch S. Measure the total current and voltage and current of each resistor in the series-parallel circuit and record the values in Table 6-1.

6. Using *Ohm's law*, calculate the total resistance of the series-parallel circuit. Record this value in Table 6-1.

7. Calculate the voltage and current of each resistor using the *rated* values of resistance and record these values in Table 6-1.

PART B: Designing a Series-Parallel Circuit

8. With the resistors from this experiment and using the structure shown in Figure 6-2, design a series-parallel circuit that will take a current of approximately 8 mA when connected across a 20 V DC power supply. Uses *rated* values in your design and show all calculations.

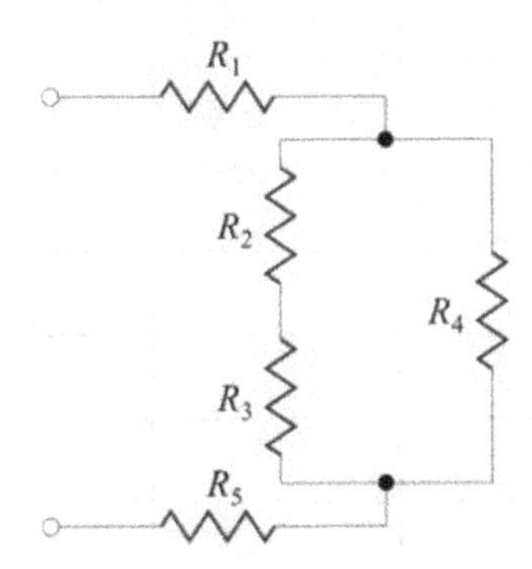

Figure 6-2

9. Draw the final circuit design showing all resistor values.

10. Construct your series-parallel circuit design, and measure the total resistance with the DMM. Record total resistance in Table 6-2.

11. Connect the power supply to the circuit. Include an ammeter for measuring total current. Adjust the applied voltage to 20 V (with a current limit of 15 mA if using a digital supply) and measure the total current flowing. Record total current in Table 6-2.

RESULTS

Table 6-1 Series-Parallel Circuit

Resistor	Resistance			Voltage		Current	
	Measured	Calculated					
		V/I	Formula	Measured	Calculated	Measured	Calculated
R_T				20 V			
330-Ω							
470-Ω							
560-Ω							
1.2-kΩ							
2.2-kΩ							
3.3-kΩ							
4.7-kΩ							

Table 6-2 Designing a Series-Parallel Circuit

Design Values			Measured Values		
E	I_T	R_T	E	I_T	R_T
20 V	8 mA				

QUESTIONS

1. In your own words, explain, the technique used to calculate the resistance of a series-parallel circuit.

2. Compare the values of total resistance from Table 6-1. Which of these values do you think is most accurate? Explain your answer.

3. When an ohmmeter is used to measure resistance in a section of a series-parallel circuit, what precautions must be taken first? Explain your answers.

4. Comment on the results of your design problem. Why was it not possible to design a circuit to draw *exactly* 8 mA?

5. Calculate the total resistance of the series-parallel circuit shown below.

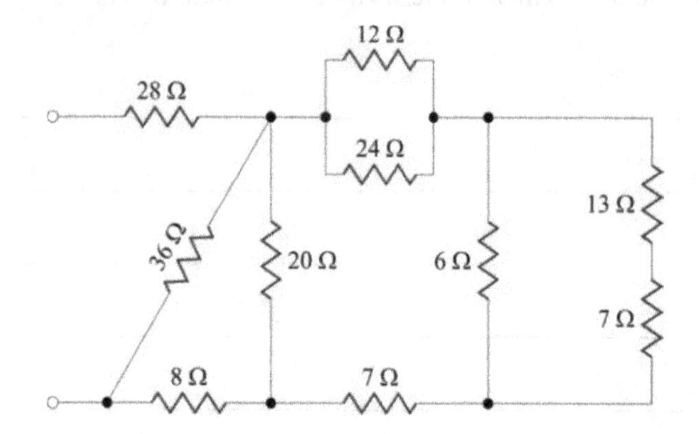

EXPERIMENT

7

THE VOLTAGE DIVIDER

OBJECTIVES

1. To verify a rule for determining voltages across resistors in an unloaded, voltage divider circuit.
2. To design a simple series-dropping circuit and test it under varying load conditions.
3. To modify the designed circuit for improved voltage regulation.

DISCUSSION

Voltage Divider Circuits

The purpose of this circuit is to reduce the supply voltage to the level required by the load.

The simplest method is the use of a series-dropping resistor.

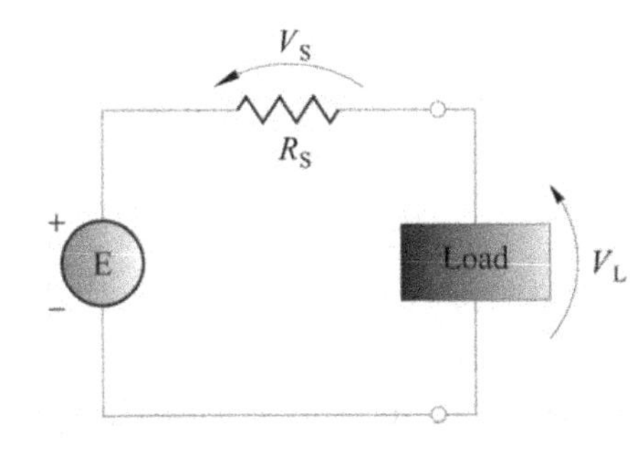

When load current flows, the series-dropping resistor causes a voltage drop (V_S) which subtracts from the supply voltage (E) giving the required voltage across the load (V_L).

From Kirchhoff's voltage law:

$$V_L = E - V_S$$

$$V_L = E - I_L R_S$$

Example: Design a voltage divider circuit to provide 6 V at a current of 0.5 amps to a motor, from a 9 V battery.

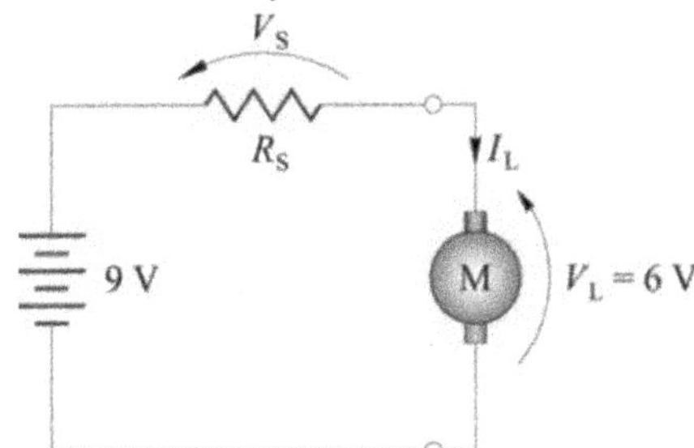

$$V_S = E - V_L = 9\text{ V} - 6\text{ V} = 3\text{ V}$$

$$R_S = \frac{V_S}{I_L} = \frac{3\text{ V}}{0.5\text{ A}} = \mathbf{6\ \Omega}$$

$$P_S = V_S I_S = 3\text{ V} \times 0.5\text{ A} = \mathbf{1.5\ W}$$ (Use a 2 W resistor)

This simple circuit has a serious disadvantage because the value of R_S is based on a particular value of load current and any change in load current will change the voltage dropped by the series resistor and consequently V_L will change also. This change in voltage with load is referred to as **regulation**. The circuit will only work well if the load current remains relatively constant.

The following graph indicates how the voltage will vary from its designed value when the load current changes.

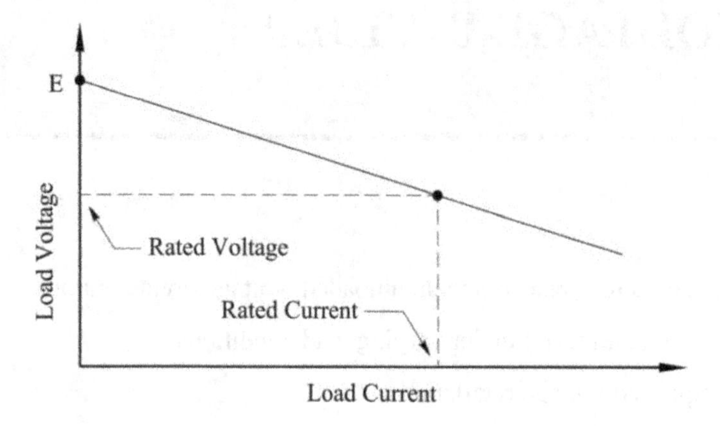

The addition of a **bleeder** resistor in parallel with the load will make the circuit less sensitive to changes in load current.

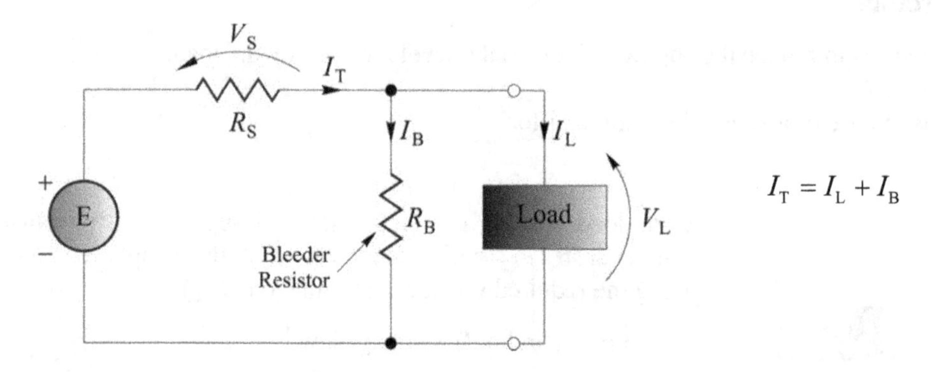

If the bleeder current is large compared to the load current, then I_T will remain more constant with changes in I_L and the load voltage will have better regulation.

Voltage regulation is the percent change in load voltage as the load is varied.

$$\%\text{Regulation} = \frac{V_1 - V_2}{V_1} \times 100\%$$

V_1 = Voltage at designed load current

V_2 = Voltage at new load current

EQUIPMENT

Power Supplies:
- Variable 0-24V DC, regulated

Instruments:
- Digital Multimeter (DMM)

Resistors: (5%, ½ W or higher)
- One each of 560-Ω, 1-kΩ, 1.5-kΩ, 2.7-kΩ, and 3.3-kΩ

Resistors: 5W
- 500-Ω

Miscellaneous:
- 5-kΩ potentiometer

PROCEDURE

PART A: Unloaded Voltage Divider

1. Connect the circuit shown below in Figure 7-1. Adjust the source voltage to 20 V (with a current limit of 10 mA if using a digital supply).

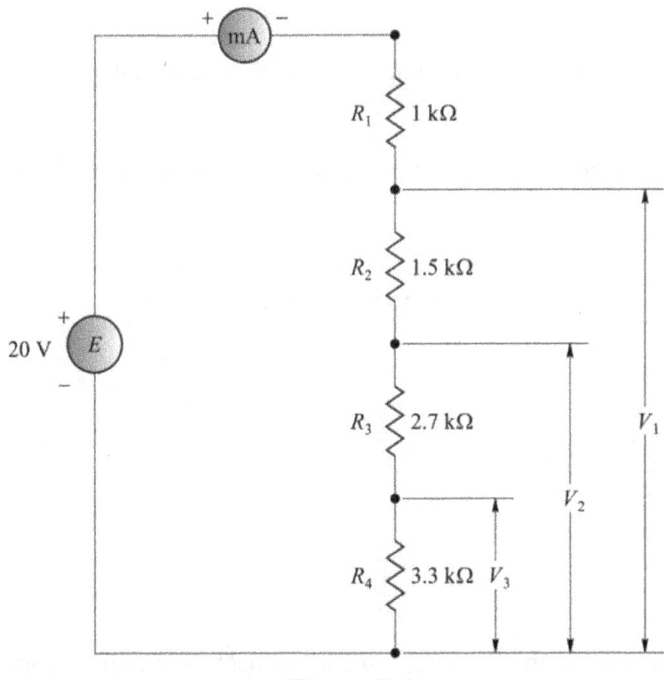

Figure 7-1

2. Turn **on** the power supply output and measure the voltage across each resistor. Record these values in Table 7-1.

3. Measure voltages V_1, V_2, and V_3 and current I. Record your measurements in Table 7-1.

4. Using circuit analysis, calculate the voltages and current and record the results in Table 7-1.

PART B: Voltage Divider Design and Regulation

5. Design a series voltage divider to deliver 15 V from a 20-V power supply to a 1-kΩ load resistor, as shown in Figure 7-2. Record the designed value of series resistance in Table 7-2.

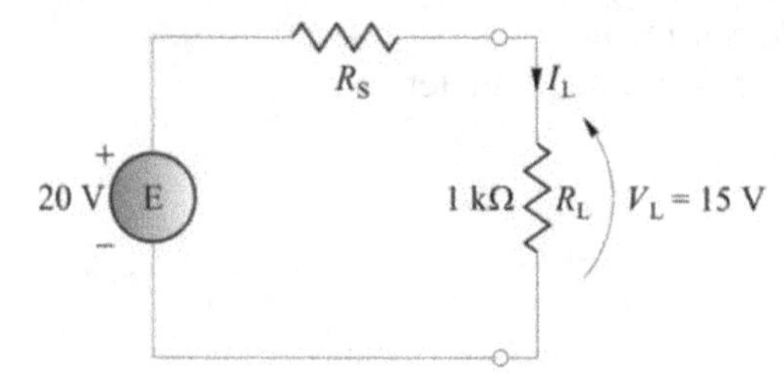

Figure 7-2

6. Set the 5-kΩ potentiometer to the value calculated for R_S in step 5 and connect the circuit as shown in Figure 7-2. Set the supply voltage to 20 V (with a current limit of 50 mA if using a digital supply).

7. Measure and record the voltage across the load resistance and the current through the load resistance. Record your results in Table 7-2.

8. Replace the load resistance with a 560-Ω resistor and measure the new value of load voltage and load current.

9. Calculate the % regulation (see discussion for formula) from the voltage at the designed load current to the voltage at the new load current.

10. Redesign the circuit using a bleeder resistor as shown in Figure 7-3. Design for a bleeder current of 30 mA. Record your designed values of series resistance and bleeder resistance in Table 7-2.

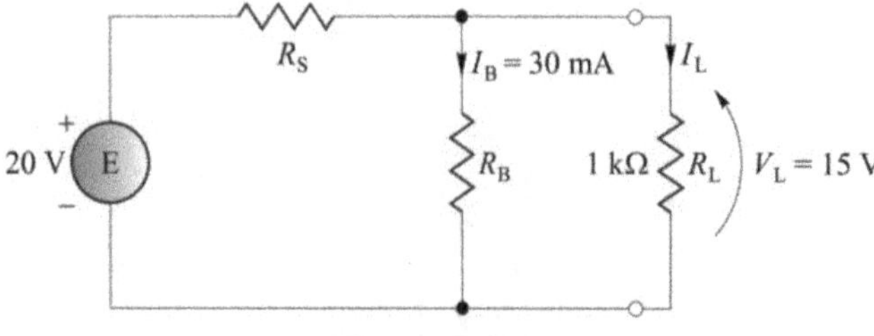

Figure 7-3

11. Set the 5-kΩ potentiometer to the value calculated for R_S in step 10 and connect the circuit as shown in Figure 7-3. Set the supply voltage to 20 V (with a current limit of 50 mA if using a digital supply).

12. Measure and record the voltage across the load resistance and the current through the load resistance. Record your results in Table 7-2.

13. Replace the load resistance with a 560-Ω resistor and measure the new value of load voltage and load current.

14. Calculate the % regulation from the voltage at the designed load current to the voltage at the new load current.

RESULTS

Table 7-1 Unloaded Voltage Divider

	E (V)	I (ma)	V_{R1}	V_{R2}	V_{R3}	V_{R4}	V_1	V_2	V_3
Measured	20								
Calculated									

Table 7-2 Voltage Divider Design and Regulation

Circuit	Design Values				Measured Values				% Voltage Regulation
					1-kΩ Load		560-Ω Load		
	R_S (Ω)	P_S (W)	R_B (Ω)	P_B (W)	V_L	I_L	V_L	I_L	
Series →									
Bleeder →									

QUESTIONS

1. Referring to Table 7-1, compare the measured values of voltage to calculated values. Explain any discrepancies.

____

____

____

____

____

2. How does the measured value of V_L in step 7 compare with the designed value? Explain any discrepancies.

____

____

____

____

____

____

3. What is the effect on load voltage when the load resistance changes in step 8? Explain why this happens.

____

____

____

____

____

____

____

____

4. Compare the voltage regulation for the two designed voltage divider circuits. Explain why the circuit with the bleeder resistor has better regulation.

5. A voltage divider supplies a load with 30 V at a current of 500 mA from a 50-V supply. If the supply current is 800 mA, determine the required value of a series-dropping resistor.

EXPERIMENT

8

POWER IN DC CIRCUITS AND MAXIMUM POWER TRANSFER

OBJECTIVES

1. To determine power in a DC circuit using various formulae.
2. To investigate the theory of maximum power transfer.

DISCUSSION

Power in DC Circuits

In mechanics, **work** is done when a force causes an object to move.

$$\text{work} = \text{force} \times \text{distance}$$

The SI unit of work is the **joule.** The English unit is the **foot-pound**.

Power is defined as the *rate* of doing work, i.e., how rapidly the work is done.

$$power = \frac{work}{time} = \frac{w}{t}$$

The unit of power is the **watt** (joule/s).

Electrical Power

In an electric circuit, the force is the source of EMF and work is done as electrons flow through the circuit. Therefore, electrical power is being delivered to the circuit.

P(electrical) = electrical force x rate of electron flow

$$P = E \times I$$

The unit of electrical power is still the **watt**.

Example 1: What is the power consumed by a 120-V lamp if it takes a current of 500 mA?

$$P = E\,I = 120\,\text{V} \times 0.5\,\text{A} = \mathbf{60\,W}$$

The power formula may be combined with Ohm's law to express power in terms of resistance.

$$P = E\,I \quad \cdots\cdots 1 \quad \text{and} \quad I = \frac{E}{R} \quad \cdots\cdots 2$$

Substitute equation 2 in equation 1 ⇨ $P = E \times \frac{E}{R} = \frac{E^2}{R}$

Also $E = I\,R$ ⇨ $P = (I\,R) \times R = I^2 R$

Summarizing $P = E\,I = \frac{E^2}{R} = I^2\,R$

Example 2: An electric toaster has a resistance of 10-Ω and operates from a 120-V supply. What power does it consume?

$$P = \frac{E^2}{R} = \frac{(120\ \text{V})^2}{10\ \Omega} = \mathbf{1440\ W}$$

Maximum Power Transfer

When a source delivers power to a load, the maximum power transferred will occur when the resistance of the load is equal to the internal resistance of the source. Consider the following circuit:

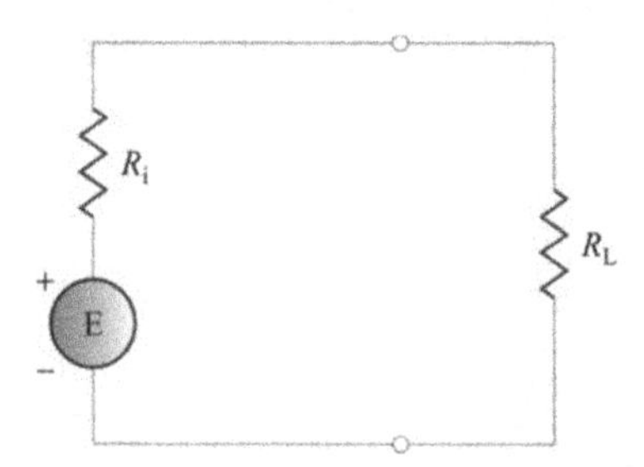

In the above circuit, R_i represents the internal resistance of the power supply.

The power consumed by the load resistance R_L is given by ⇨ $P_L = I_L^{\,2}\,R_L$

The current is calculated from Ohm's law ⇨ $I_L = \frac{E}{R_T}$

Substituting this equation into the previous formula, the load power (P_L) can be written as

$$P_L = \left(\frac{E}{R_T}\right)^2 \times R_L = \frac{E^2\,R_L}{(R_i + R_L)^2}$$

As the load resistance (R_L) is varied, the load power (P_L) will increase to a maximum and then decrease again. The following graph results if P_L is plotted against R_L.

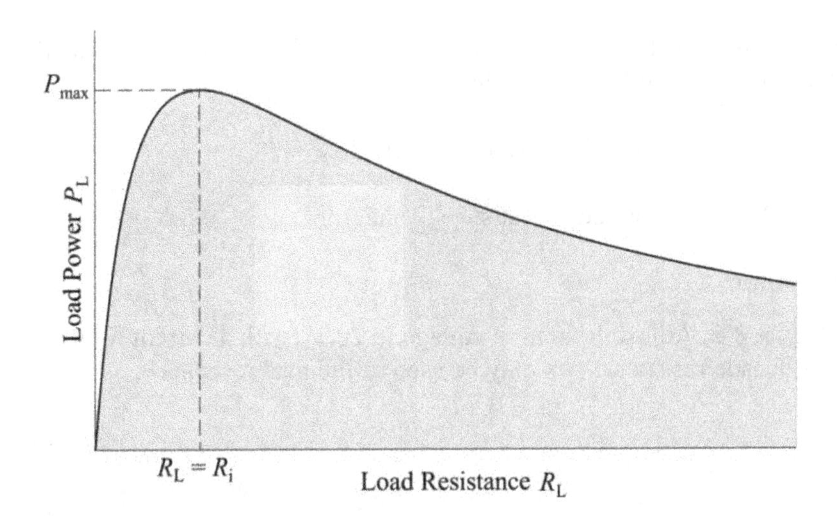

Maximum power will be transferred from the source to the load when the value of load resistance is equal to the source resistance.

This is important in such applications as communications and amplifiers where the resistance of the speaker (load) must be equal to the output resistance of the amplifier for maximum power to be delivered.

EQUIPMENT

Power Supplies:

- Variable 0-24 V DC, regulated

Instruments:

- Digital Multimeter (DMM)

Resistors: (5% ½ W or higher)

- One each of 47-Ω, 100-Ω, 220-Ω, 330-Ω, 470-Ω, and 560-Ω
- One each of 1.5 kΩ, 2.2-kΩ, 3.3-kΩ, 4.7-kΩ, 5.6-kΩ, 8.2-kΩ, and 10-kΩ
- Two each of 1-kΩ and 1.2-kΩ
- Decade resistance box (optional)

PROCEDURE

PART A: Power in a DC circuit

1. Connect the circuit shown in Figure 8-1. R_i represents the internal resistance of the source.

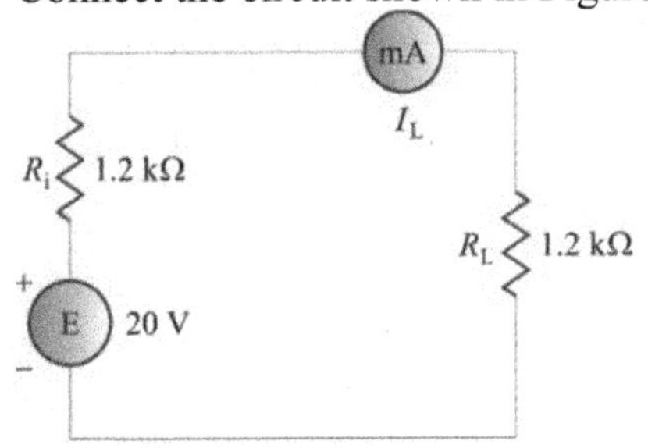

Figure 8-1

2. Turn **on** the power supply and adjust the source voltage to 20 V (with a current limit of 25 mA if using a digital supply). Measure the voltage across the load resistor and the current through the load resistor. Record these values in Table 8-1.

3. Using the three different power formulas, calculate the power delivered to the load resistor. Calculate the total power delivered by the source. Record the values in Table 8-1.

4. Replace the load resistor with a 330-Ω resistor and repeat steps 2 and 3.

5. Replace the load resistor with a 4.7-kΩ resistor and repeat steps 2 and 3.

PART B: Maximum Power Transfer

6. Connect the circuit shown in Figure 8-2. Adjust the source voltage to 20 V (with a current limit of 25 mA if using a digital supply). A decade resistance box may be used as the load resistance.

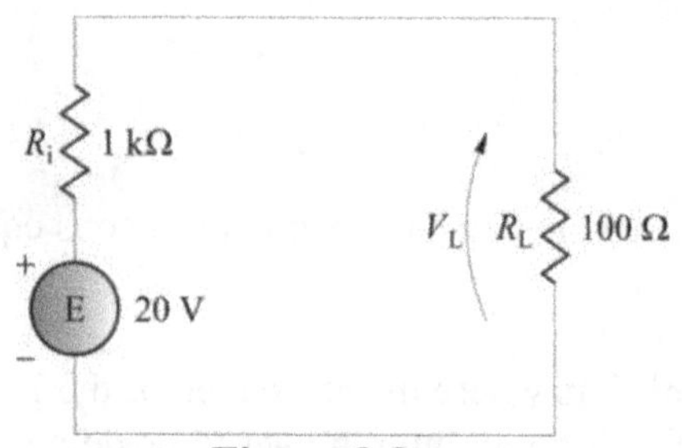

Figure 8-2

7. Measure the voltage (V_L) across the load resistance. Record the value in Table 8-2.

8. Repeat step 7 for each of the load resistance values shown in Table 8-2. Make sure the source voltage remains at 20 V for each step.

9. Using the measured values of V_L and the rated value of R_L, calculate the power delivered to the load for each different load resistance. Use the formula: $P_L = \frac{V_L^2}{R_L}$. Record your values in Table 8-2.

10. Calculate the total power delivered by the power supply for each load step using the following formula: $P_T = \frac{E^2}{R_T}$ Record your values in Table 8-2.

11. Calculate the efficiency of power transfer (η) and record the values in Table 8-2.

12. On the sheet of graph paper provided, using the same axes, plot graphs of P_L vs. R_L and P_T vs. R_L. Plot R_L on the horizontal axis.

RESULTS

Table 8-1 Power in a DC circuit

Load R_L	E (V)	V_L (V)	I_L (mA)	Load Power (W)			Total Power (W)
				$V_L I_L$	$I_L^2 R_L$	V_L^2 / R_L	
1.2-kΩ	20						
330-Ω	20						
4.7-kΩ	20						

Table 8-2 Maximum Power Transfer

R_i (Ω)	R_L (Ω)	V_L (V)	$P_L = \frac{V_L^2}{R_L}$ (W)	$P_T = \frac{E^2}{R_T}$ (W)	$\eta = \frac{P_L}{P_T}$
1000	0				
1000	100				
1000	220				
1000	330				
1000	470				
1000	560				
1000	1000				
1000	1500				
1000	2200				
1000	3300				
1000	4700				
1000	8200				
1000	10 000				

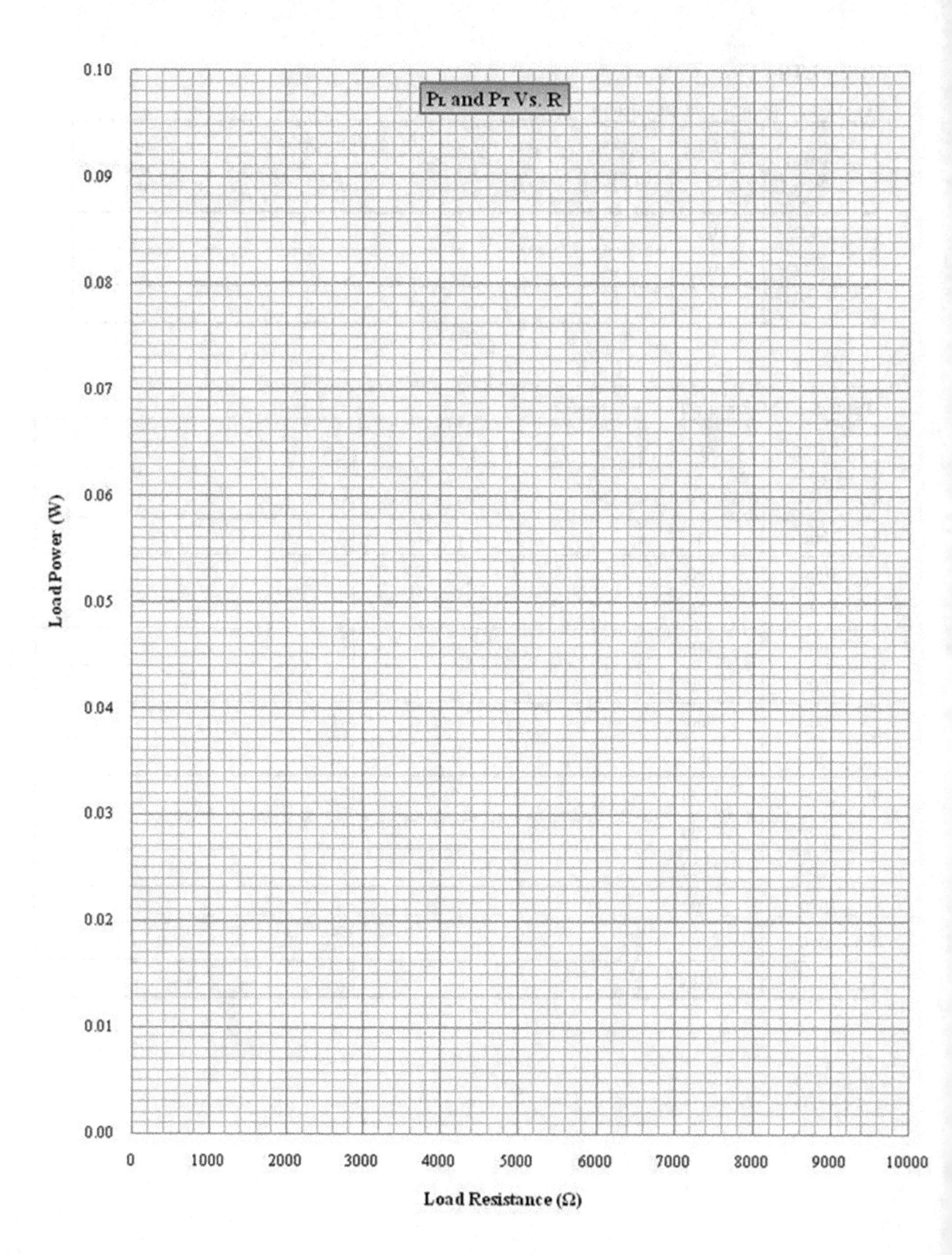

P_L and P_T Vs. R
Load Power (W)
0.10
0.09
0.08
0.07
0.06
0.05
0.04
0.03
0.02
0.01
0.00
0
1000
2000
3000
4000
5000
6000
7000
8000
9000
10000
Load Resistance (Ω)

QUESTIONS

1. Referring to the 1.2-kΩ load resistor in Table 8-1, compare the values of power calculated by each formula. Should they be the same? Explain any discrepancies.

2. Referring to Table 8-1, compare the load power for the three different load resistors. Explain why the 1.2-kΩ resistor has the most power.

3. In your own words, explain the relationship between load resistance and the transfer of power between source and load.

4. Show on your graph the value of R_L where maximum power occurs. Does this confirm the relationship discussed in Question 3? Explain any discrepancies.

5. A 50-V source has an internal resistance of 550 Ω. What is the maximum power this source can deliver to a load?

EXPERIMENT

9

MESH ANALYSIS

OBJECTIVE

1. To verify mesh analysis for experimental circuits.

DISCUSSION

Mesh Analysis

This technique is used to write voltage equations by simple inspection of the circuit, without detailed analysis.

Consider the following general circuit:

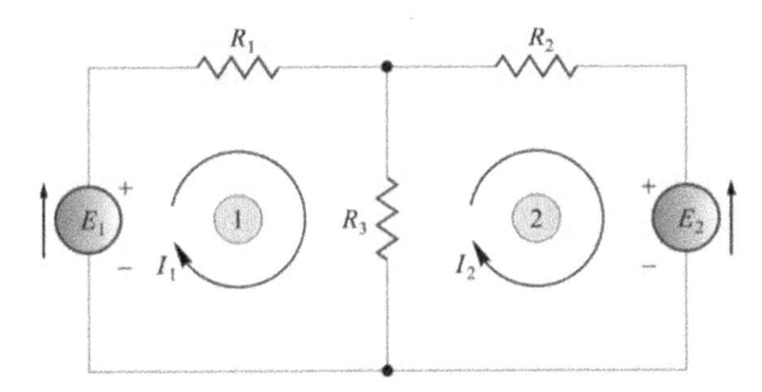

Mesh: *A loop containing no other loops*. There are 3 loops but only 2 meshes.

I_1 and I_2 are called *mesh* currents and are the currents flowing in the particular meshes. They are always taken in a clockwise direction.

These are the unknown currents in the circuit and all other currents can be expressed in terms of these. For example, the current in $R_2 = I_1 - I_2$ (Kirchhoff's current law).

General Equation for N Mesh Network

For mesh N

$$R_{NN}I_N - R_{N1}I_1 - R_{N2}I_2 - \cdots\cdots\cdots = E_N$$

R_{NN} - The sum of the resistances in mesh N
R_{N1} - Resistance common to meshes N and 1
R_{N2} - Resistance common to meshes N and 2
E_N - Algebraic sum of source voltages in mesh N.

Example: For the circuit shown, write the mesh equations and find the current in the 2-Ω resistor.

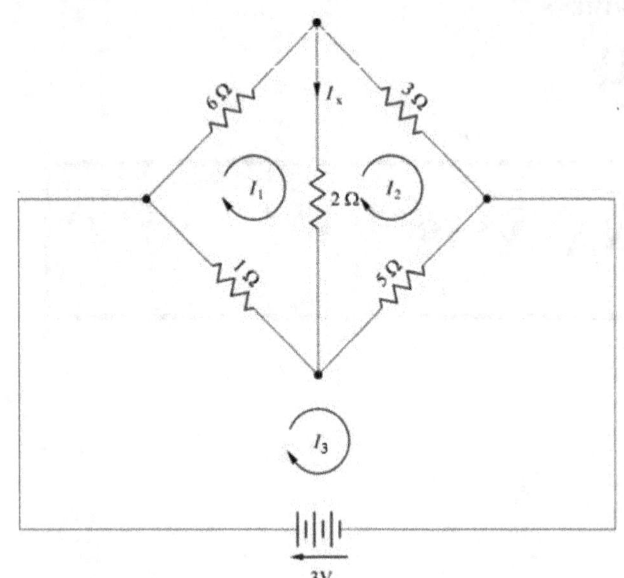

$I_x = I_1 - I_2$

For mesh 1

$9I_1 - 2I_2 - I_3 = 0$

For mesh 2

$10I_2 - 2I_1 - 5I_3 = 0$

For mesh 3

$6I_3 - I_1 - 5I_2 = 3$

Rearranging:

$$\begin{aligned} 9I_1 - 2I_2 - I_3 &= 0 \\ -2I_1 + 10I_2 - 5I_3 &= 0 \\ -I_1 - 5I_2 + 6I_3 &= 3 \end{aligned}$$

$$I_1 = \frac{\begin{vmatrix} 0 & -2 & -1 \\ 0 & 10 & -5 \\ 3 & -5 & 6 \end{vmatrix}}{\begin{vmatrix} 9 & -2 & -1 \\ -2 & 10 & -5 \\ -1 & -5 & 6 \end{vmatrix}} = \frac{(0+30+0)-(-30+0+0)}{(540-10-10)-(10+24+225)} = \frac{60}{261} = 0.230\,\text{A}$$

$$I_2 = \frac{\begin{vmatrix} 9 & 0 & -1 \\ -2 & 0 & -5 \\ -1 & 3 & 6 \end{vmatrix}}{D} = \frac{(0+0+6)-(0+0-135)}{261} = \frac{141}{261} = 0.540\,\text{A}$$

$I_x = I_1 - I_2 = 0.230\text{ A} - 0.540\text{ A} =$ **-0.310 A** (assumed in the wrong direction)

EQUIPMENT

Power Supplies:
- Two variable 0-24 V DC, regulated

Instruments:
- Digital Multimeter (DMM)

Resistors: (5%, ½ W or higher)
- One each of 1-kΩ, 1.5-kΩ, 2.2-kΩ, 3.3-kΩ, 5.6-kΩ, and 6.8-kΩ

PROCEDURE

1. Connect the circuit shown in Figure 9-1.

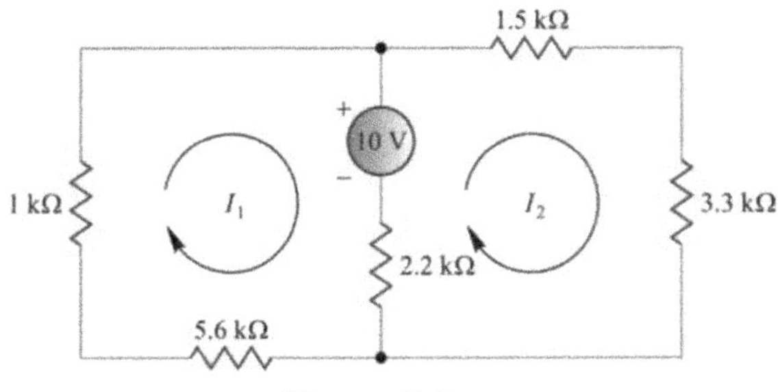

Figure 9-1

2. Set the power supply to 10 V (with a current limit of 15 mA if using a digital supply).

3. Measure the current in each mesh and in the 2.2-kΩ resistor. Record your results in Table 9-1.

4. Calculate theoretical values and record in Table 9-1. Show all calculations.

5. Connect the circuit shown in Figure 9-2.

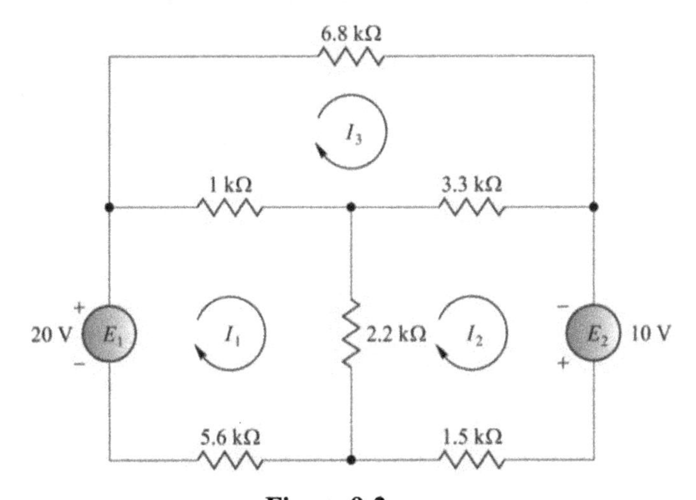

Figure 9-2

6. Set one power supply to 20 V and a second power supply to 10 V (each with a current limit of 15 mA if using digital supplies).

7. Measure the mesh currents and the currents in the common resistors. Record the values in Table 9-2.

8. Calculate the theoretical values and record in Table 9-2. Show all calculations.

RESULTS

Table 9-1 Currents: Step 3

Current (ma)	I_1	I_2	$I_{2.2k}$
Measured			
Calculated			
% Error			

Table 9-2 Currents: Step 7

Current (mA)	I_1	I_2	I_3	I_{1k}	$I_{3.3k}$	$I_{2.2k}$
Measured						
Calculated						
% Error						

QUESTIONS

1. How do your measured values compare with calculated values? Explain any discrepancies.

2. What is the significance of a negative sign, if any, in your calculated values?

3. Upon which laws is the mesh analysis technique based?

4. For the circuit shown, write the mesh equations and find I_x.

2 Ω 9 Ω 4 Ω I_x 5 Ω 1 Ω 50 V 20 V 30 V 15 Ω 3 Ω 20 V

EXPERIMENT

10

THE SUPERPOSITION THEOREM

OBJECTIVES

1. To verify the superposition theorem for experimental circuits.

DISCUSSION

Superposition Theorem: *For any linear passive network with more than one source, the voltage or current of any particular circuit component may be found by considering each source separately and determining the current or voltage due to that source alone. The total current or voltage is the algebraic sum of the individual currents or voltages.*

This technique can be used to solve circuits containing both current and voltage sources.

When determining the values due to each source, the other sources are replaced by their internal resistances.

Voltage source = 0 Ω (short circuit)

Current source = ∞ Ω (open circuit)

Example: Find I_X by superposition.

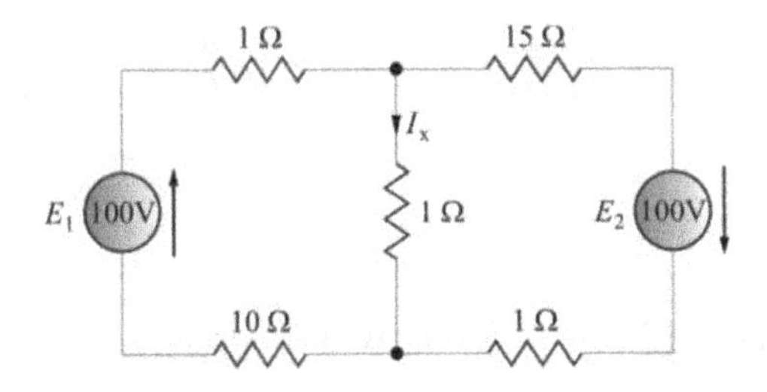

Current due to E_1 alone

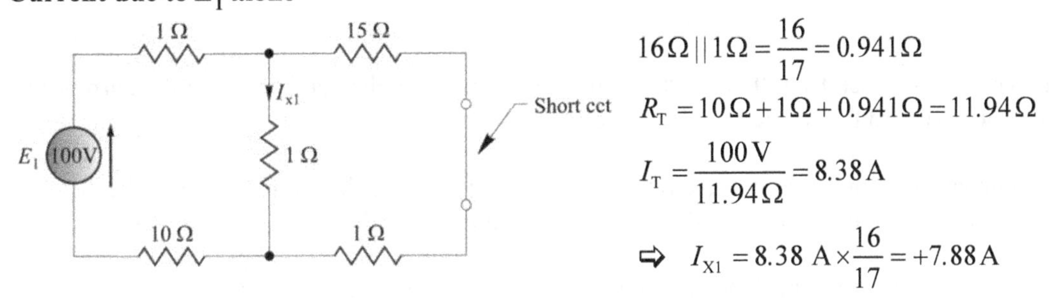

Current due to E_2 alone

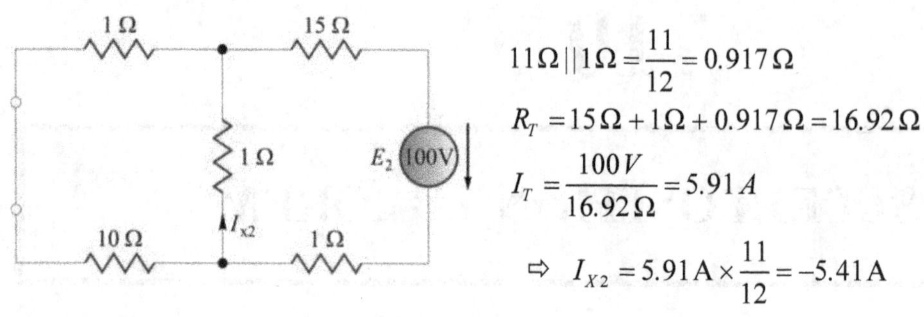

$$11\Omega || 1\Omega = \frac{11}{12} = 0.917\,\Omega$$

$$R_T = 15\,\Omega + 1\,\Omega + 0.917\,\Omega = 16.92\,\Omega$$

$$I_T = \frac{100\,V}{16.92\,\Omega} = 5.91\,A$$

$$\Rightarrow\ I_{X2} = 5.91\,\text{A} \times \frac{11}{12} = -5.41\,\text{A}$$

$$I_X = I_{X1} + I_{X2} = 7.88\,\text{A} + (-5.41\,\text{A}) = \mathbf{2.47\,A}$$

EQUIPMENT

Power Supplies:
- Two variable 0-24 DC, regulated

Instruments:
- Digital Multimeter (DMM)

Resistors: (5%, ½ W or higher)
- One each of 1-kΩ, 1.2-kΩ, 2.7-kΩ, 3.3-kΩ, 4.7-kΩ, and 10-kΩ

Miscellaneous:
- Two SPST switches

PROCEDURE

1. Connect the circuit shown in Figure 10-1.

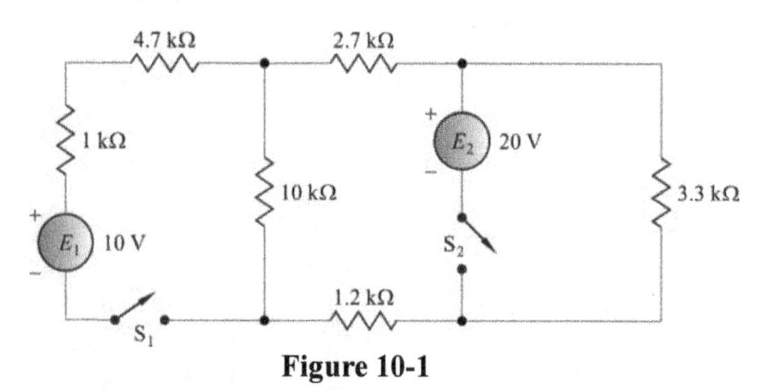

Figure 10-1

2. Adjust E_1 and E_2 to the indicated voltages (with current limits of 15 mA if using digital supplies).

3. **Close** S_1 and S_2. Measure the current in the 10-kΩ resistor with an ammeter. Record this value (I_T) in Table 10-1.

4. **Open** both switches. Replace E_2 with a short circuit. Close both switches and measure the current in the 10-kΩ resistor. Record this value (I_{X1}) in Table 10-1.

5. **Open** both switches. Reconnect E_2 in the circuit and replace E_1 with a short circuit. Close the switches and measure the current in the 10-kΩ resistor. Record this value (I_{X2}) in Table 10-1.

6. Reverse the polarity of E_2 and repeat steps 3 to 5. Record your results in Table 10-1.

7. Connect the circuit shown in Figure 10-2.

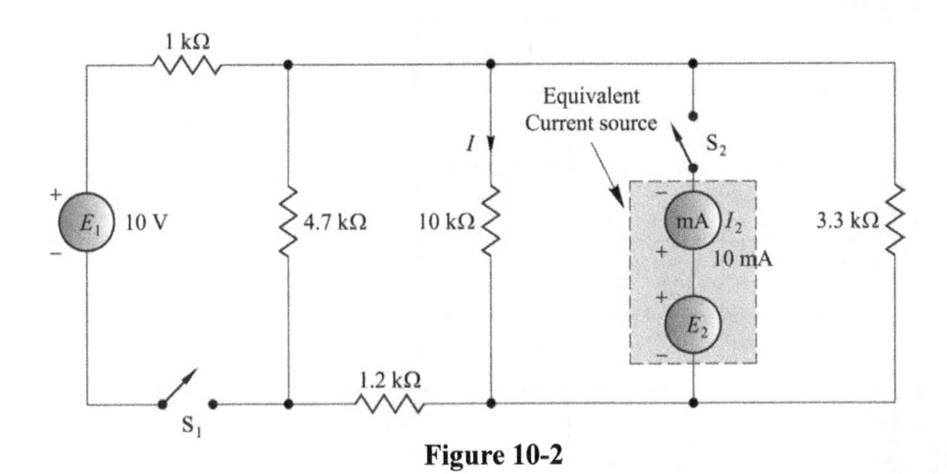

Figure 10-2

8. **Close** S_1 and S_2. Adjust E_2 until the ammeter (I_2) indicates 10 mA. This is now equivalent to a constant current source of 10 mA.

9. Measure the current (I_T) flowing in the 10-kΩ resistor and record in Table 10-2.

10. Replace the current source with an open circuit (i.e., open switch S_2). Measure and record the current in the 10-kΩ resistor (I_{x1}).

11. **Open** both switches. Replace E_1 with a short circuit. **Close** both switches and readjust I_2 to 10 mA. Measure and record the current in the 10-kΩ resistor (I_{x2}).

12. Reverse the polarity of E_1 and repeat steps 8 to 11. Record results in Table 10-2.

13. Calculate theoretical values for both circuits. Show all calculations.

RESULTS

Table 10-1 Currents Figure 10-1

Current (mA)	I_T	I_{X1}	I_{X2}	$I_{X1} + I_{X2}$
Steps 3-5				
Calculated				
% Error				
Step 6				
Calculated				
% Error				

Table 10-2 Currents Figure 10-2

Current (mA)	I_T	I_{X1}	I_{X2}	$I_{X1} + I_{X2}$
Steps 9-11				
Calculated				
% Error				
Step 12				
Calculated				
% Error				

QUESTIONS

1. In your own words, describe how the superposition theorem is used to solve a multi-source network.

---

2. How do your measured values compare with calculated values?

---

3. What are the principal sources of error in this experiment?

---

4. What effect did reversing the polarity of the power supply have on I_T in each circuit? Explain your answer.

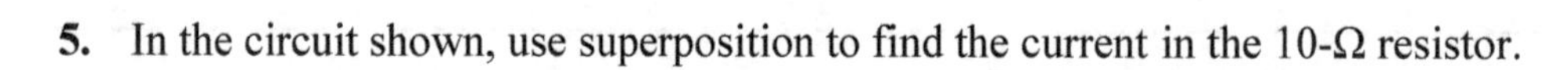

5. In the circuit shown, use superposition to find the current in the 10-Ω resistor.

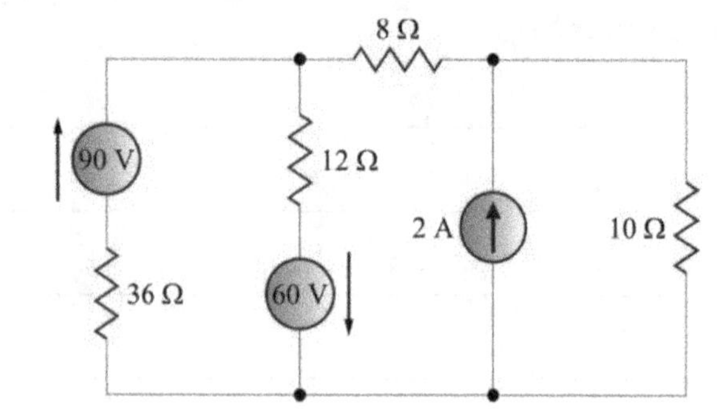

EXPERIMENT

11

THÉVENIN'S THEOREM

OBJECTIVES

1. To determine the Thévenin-equivalent circuit of a complex DC circuit.
2. To verify the validity of Thévenin's theorem for varying load resistances.

DISCUSSION

Thévenin's Theorem

An electrical network is contained in a box with two terminals brought out. Using only a voltmeter and ammeter we try to determine the circuitry inside the box.

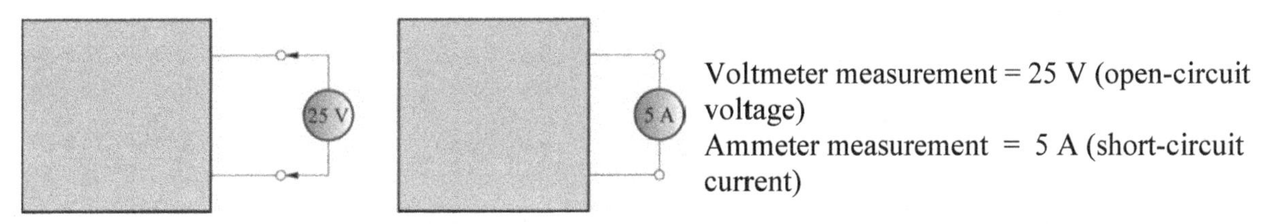

Voltmeter measurement = 25 V (open-circuit voltage)
Ammeter measurement = 5 A (short-circuit current)

This indicates that $R = \frac{V}{I} = \frac{25\,\text{V}}{5\,\text{A}} = 5\,\Omega$

The circuit appears to consist of a 25-V source in series with a 5-Ω resistor.

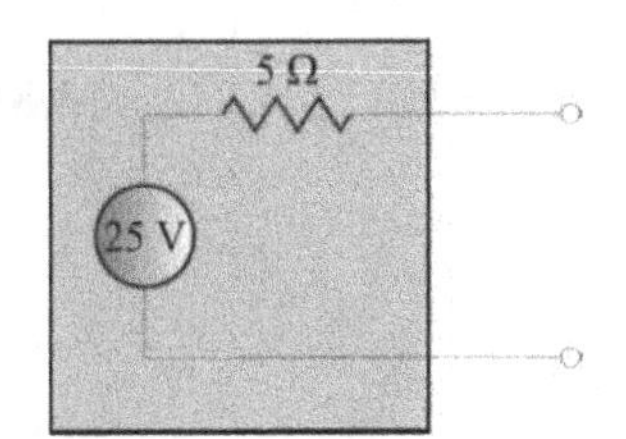

Connecting a 5-Ω and a 20-Ω resistor between the terminals and measuring the current supports this assumption.

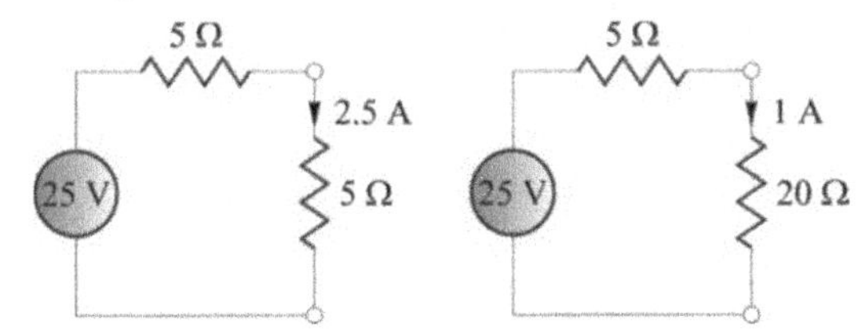

When the box is opened the following circuit is found inside:

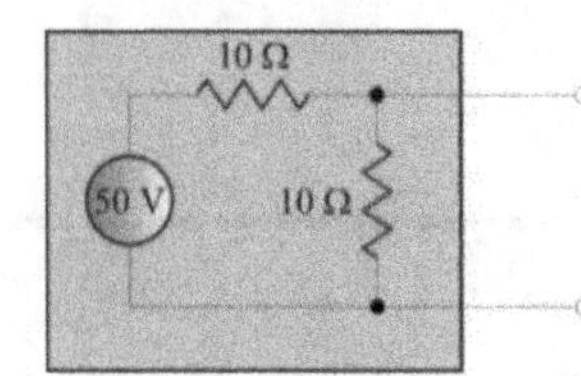

The circuit we determined has the same voltage and current for all resistances connected between the external terminals as the actual circuit in the box. It is therefore an equivalent circuit.

Thévenin's theorem involves the replacement of a complex network by a simple series-circuit with respect to two terminals in the network called the **load terminals**. The current calculated in this series circuit is the same as the current flowing in the original circuit for all conditions of load resistance.

The Thévenin-equivalent circuit consists of a voltage source (Thévenin source), which is the open circuit voltage across the load points, in series with a resistance (impedance). This is the resistance between the load points with the load removed and all sources replaced by their internal resistances.

Example: For the circuit shown below, calculate the current in the 2-Ω resistor.

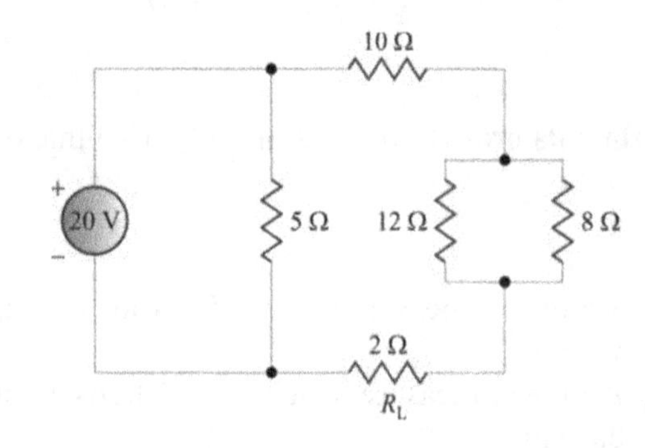

Step 1. Find E_{Th} (open-circuit voltage) Step 2. Find R_{TH} (Thévenin resistance)

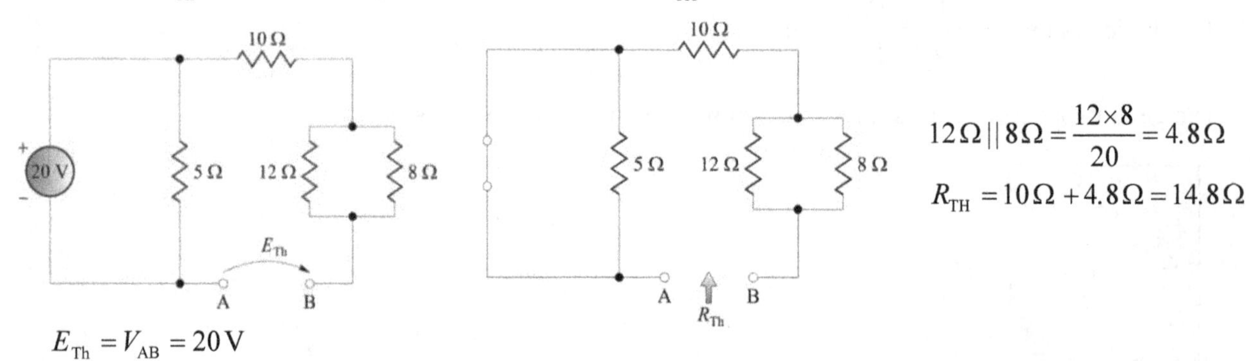

$$12\Omega \,||\, 8\Omega = \frac{12\times 8}{20} = 4.8\Omega$$

$$R_{TH} = 10\Omega + 4.8\Omega = 14.8\Omega$$

$$E_{Th} = V_{AB} = 20\,\text{V}$$

Thevenin-equivalent Circuit

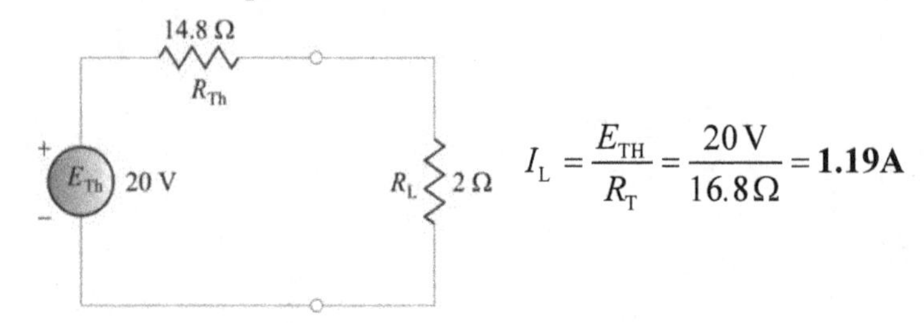

$$I_L = \frac{E_{TH}}{R_T} = \frac{20\,\text{V}}{16.8\Omega} = \mathbf{1.19A}$$

EQUIPMENT

Power Supplies:
- Variable 0-24 V DC, regulated

Instruments:
- Digital Multimeter (DMM)

Resistors: (5%, ½ W or higher)
- One each of 1-kΩ, 1.2-kΩ, 2.2-kΩ, 2.7-kΩ, 3.3-kΩ, 4.7-kΩ, 6.8-kΩ, and 10-kΩ

Miscellaneous:
- 5-kΩ potentiometer

PROCEDURE

1. Using circuit analysis techniques, determine the Thévenin-equivalent circuit for the network external to the 10-kΩ resistor in Figure 11-1. Record the values in Table 11-1. Write the values of E_{Th} and R_{Th} on the Thévenin-equivalent circuit shown.

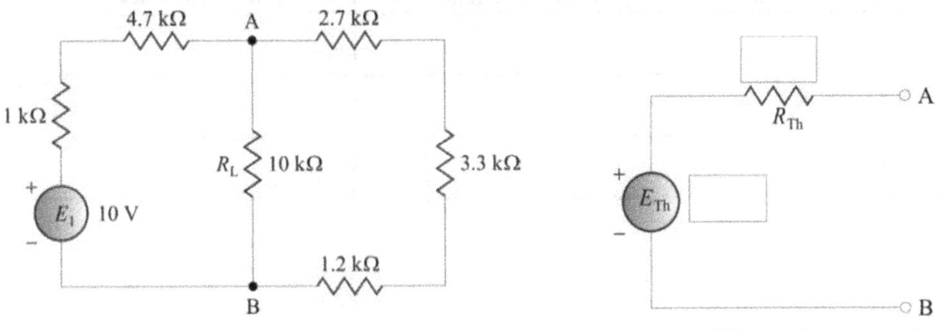

Figure 11-1 **Thevenin-equivalent Circuit**

2. Connect the circuit shown in Figure 11-1. Remove the 10-kΩ load resistor (R_L) from the circuit. Switch **on** the power supply and adjust it to 10 V (with a current limit of 10 mA if using a digital supply). Measure the voltage across the open circuit load points (A-B). This is the Thévenin-equivalent voltage, E_{Th}. Record this value in Table 11-1.

3. Remove source E_1 and replace it with a short circuit. Measure the resistance across the open-circuit load points (A-B). This is the Thévenin-equivalent resistance, R_{Th}. Record this value in Table 11-1.

4. Remove the short circuit and reconnect the power supply. Re-install the 10-kΩ load resistor. Measure the voltage across the 10-kΩ load resistance (R_L). Record your values in Table 11-2.

5. Repeat step 4 for load resistances of 6.8-kΩ and 2.2-kΩ.

6. Using the Thévenin-equivalent circuit determined in **step 1**, calculate the load voltage for a load of 10-kΩ. Repeat the calculation for loads of 6.8-kΩ and 2.2-kΩ. Record values in Table 11-2.

7. Adjust the 5-kΩ potentiometer to the value of R_{Th} calculated in step 1. Connect the Thévenin-equivalent circuit using the potentiometer and the source set to the value of E_{Th} determined in step 1. Connect the 10-kΩ load resistor and measure the voltage across it. Record your results in Table 11-2.

8. Repeat step 7 for load resistances of 6.8-kΩ and 2.2-kΩ.

RESULTS

Table 11-1 Thévenin-equivalent Circuit

Thevenin Value→	E_{Th}	R_{Th}
Calculated		
Measured		

Table 11-2 Load Voltages

Load Resistance	Load Voltage (V_L)		
	Measured in Original Circuit	Calculated in Thévenin Circuit	Measured in Thévenin Circuit
10-kΩ			
6.8-kΩ			
2.2-kΩ			

QUESTIONS

1. In your own words, explain how the Thévenin-equivalent circuit method is used to solve for voltage or current in a complex network.

2. Under what conditions does the Thévenin technique have a significant advantage over other circuit analysis techniques? Explain.

3. Compare measured voltages from the original circuit with corresponding voltages from the Thévenin circuit. What conclusion can you draw from this comparison?

4. Comment on the difficulty of calculating the load voltages from the circuit shown in Figure 11-1 versus calculating them from the Thévenin-equivalent circuit.

5. For the circuit shown below, calculate the current in the 40-Ω resistor using Thévenin's theorem.

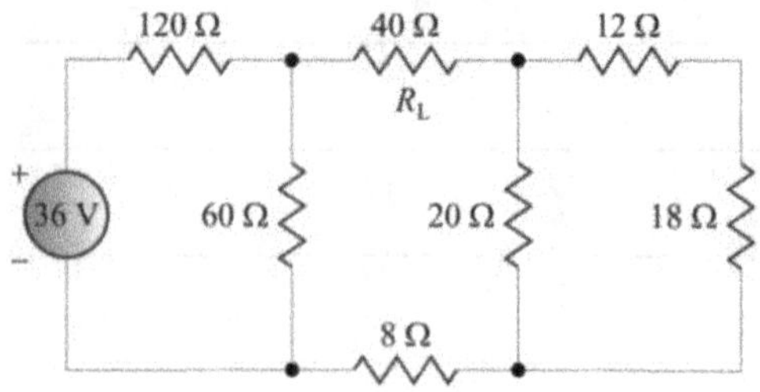

120 Ω   40 Ω   12 Ω

R_L

36 V   60 Ω   20 Ω   18 Ω

8 Ω

EXPERIMENT

12

THE DC AMMETER AND VOLTMETER

OBJECTIVES

1. To determine how a basic meter movement can have its range extended by the addition of a shunt resistor.
2. To verify the shunt resistor required for specific ranges.
3. To determine the multipliers required to convert a meter movement into a voltmeter with specific ranges.
4. To investigate the loading effect of a voltmeter.

DISCUSSION

Moving-Coil Meter Movement

Before the development of digital meters, the most common basic meter movement was the moving-coil or **D'Arsonval** meter movement. This movement can be used as a DC ammeter, DC voltmeter, or ohmmeter.

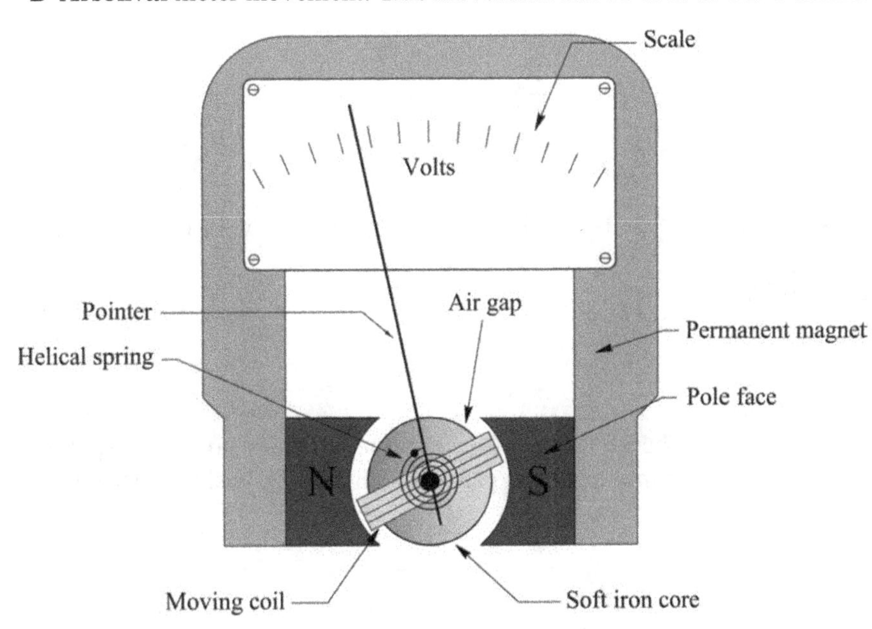

The operation of the meter is based on the motor principle. When a current flows through the coil, it becomes an electromagnet with a north and a south pole. The poles are repelled by the permanent magnet field, the coil rotates and the pointer moves upscale.

The iron core does *not* rotate but the coil rotates in the air gaps between the core and the poles of the permanent magnet.

The iron core is used to provide a low-reluctance path for the magnetic flux and therefore increase the flux in the air gap.

If the current in the coil is increased, the strength of the electromagnetic field increases, causing a stronger repulsive force and therefore a larger deflection. The amount of deflection is directly proportional to the current flowing and therefore the scale is linear.

If the current in the coil is reversed, the direction of the electromagnetic field will reverse, and the coil will rotate in the opposite direction. It is important to ensure proper polarity on the meter connections to avoid damage.

The DC Ammeter

The basic movement can be used directly as an ammeter. For example a 1-mA, 100-Ω movement can be used directly as a 0-10 mA ammeter.

1 mA

To extend the range of the ammeter, a resistor called a **shunt** can be connected in parallel with the movement. This resistor will divert most of the current, allowing only a maximum of rated current to actually flow through the movement.

$$R_{sh} = \frac{V_{sh}}{I_{sh}}$$

$$V_{sh} = V_m = I_m R_m$$

$$I_{sh} = I_T - I_m \quad \Rightarrow \quad R_{sh} = \frac{I_m R_m}{I_T - I_m}$$

Example: Determine the shunt resistance to convert a 1-mA, 100-Ω movement into a 0-10 mA ammeter.

$$R_{sh} = \frac{0.001\,\text{A} \times 100\,\Omega}{0.01\,\text{A} - 0.001\,\text{A}} = \frac{0.1\,\text{V}}{0.009\,\text{A}} = \mathbf{11.11\,\Omega}$$

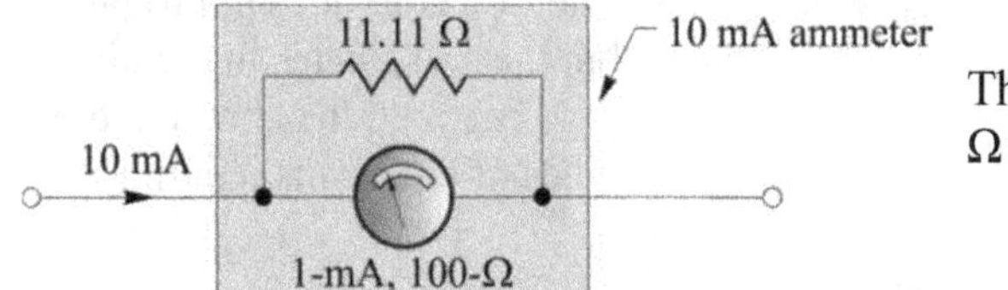

Therefore, the 10-mA ammeter consists of a 1-mA, 100-Ω movement in parallel with a 11.11-Ω shunt resistor.

The DC Voltmeter

The same moving-coil movement can be used as a DC voltmeter. Using the 1-mA, 100-Ω movement, at full-scale deflection, $I = 1$ mA and the corresponding voltage drop is $V = 0.001\,\text{A} \times 100\,\Omega = 0.1\,\text{V}$. Therefore, this movement by itself could only measure a maximum voltage of 0.1 V, which has no practical application.

To extend the range of the voltmeter, a resistance called a **multiplier** can be connected in series. This resistor will drop most of the circuit voltage, allowing only a maximum voltage of 0.1 V across the movement.

Example: Calculate the multiplier resistance required to convert the movement into a 0-20 V voltmeter.

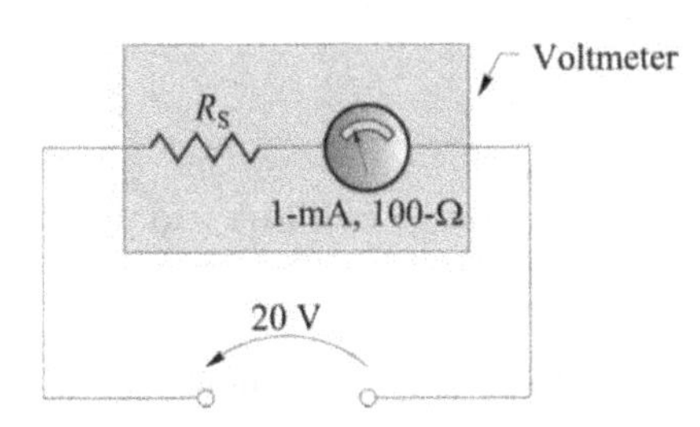

$$R_T = \frac{V_T}{I} = \frac{20\,\text{V}}{1\,\text{mA}} = 20\,\text{k}\Omega$$

$$R_S = R_T - R_m = 20\,\text{k}\Omega - 100\,\Omega = \mathbf{19.9\,k\Omega}$$

Loading Effect of a Voltmeter

A voltmeter must have a high resistance so that it will not change the circuit significantly when the meter is connected to make a measurement. If the voltmeter resistance is not high enough compared to circuit resistance, it will cause an error by affecting the circuit and changing the voltage that it is measuring. This effect is referred to as the **loading effect**. The following numerical example demonstrates the error introduced by a low resistance voltmeter.

Example: A 1000-Ω voltmeter is used to measure the voltage across R_2 in the following circuit.

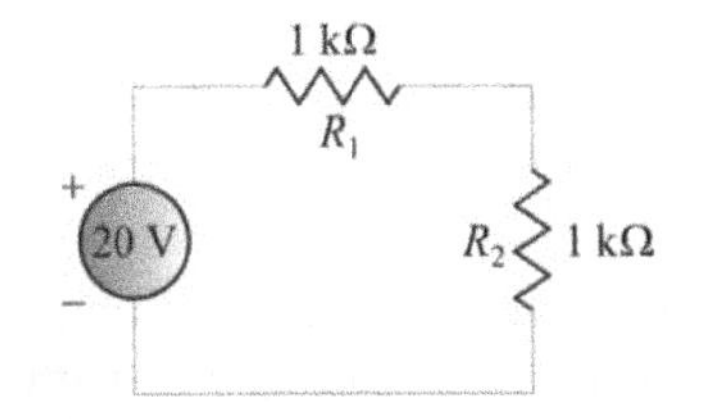

The actual voltage across $R_2 = 10$V.

With meter

1 kΩ R_1 20 V R_2 1 kΩ V R_m = 1 kΩ

$$R_2 \,||\, R_m = 1\,\text{k}\Omega \,||\, 1\,\text{k}\Omega = 500\,\Omega$$

$$V_{R_2} = 20\,\text{V} \times \frac{500\,\Omega}{1500\,\Omega} = 6.67\,\text{V} \quad \text{(Indicated on meter)}$$

$$\%\ \text{Error} = \frac{6.67\,\text{V}}{20\,\text{V}} = \mathbf{33.3\%}$$

EQUIPMENT

Power Supplies:

- ➢ Variable 0-24 V DC, regulated

Instruments:

- ➢ Digital Multimeter (DMM)
- ➢ 1-mA, 100-Ω meter movement

Resistors: (5%, ½ W or higher)

- ➢ One each of 3.3-kΩ, 8.2-kΩ, and 10-kΩ
- ➢ Two 1-kΩ

PROCEDURE

PART A: DC Ammeter

1. Using the 1-mA, 100-Ω movement, calculate the shunt resistor required for a range of 2 mA and record the value in Table 12-1.

2. Find the resistor from your kit closest to the shunt resistor value determined in step 1. Connect the circuit shown in Figure 12-1. The DMM is used as a DC ammeter to check the range of the experimental ammeter. R_1 is a current-limiting resistor to protect the meter.

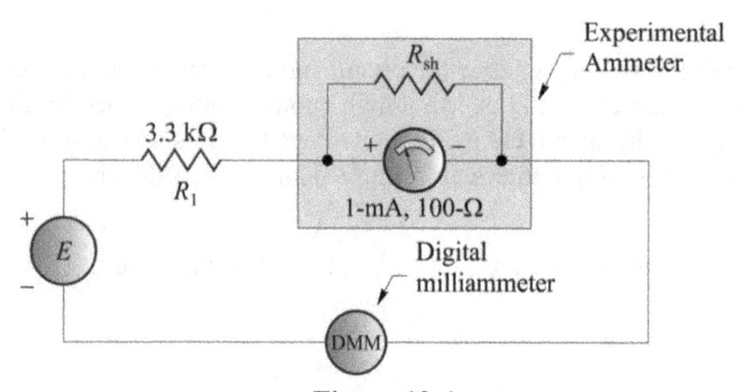

Figure 12-1

3. Turn **on** the power supply (and limit the current to 15 mA if using a digital supply). Adjust the power supply voltage for a full-scale deflection on the meter movement and measure the circuit current with the DMM. Record this value in Table 12-1.

4. Repeat this procedure for current ranges of 4 mA and 11 mA and record values in Table 12-1.

PART B: DC Voltmeter

5. Calculate the multiplier resistor required to convert the 1-mA, 100-Ω movement into a voltmeter with a range of 5 V. Record this value in Table 12-2.

6. Find the resistor from your kit closest to the multiplier value determined in step 5. Construct the experimental voltmeter by connecting the multiplier in series with the meter movement.

7. Connect the experimental voltmeter and the DMM set for voltage measurement in parallel with the DC power supply as shown in Figure 12-2. The DMM is used as a DC voltmeter to check the range of the experimental voltmeter.

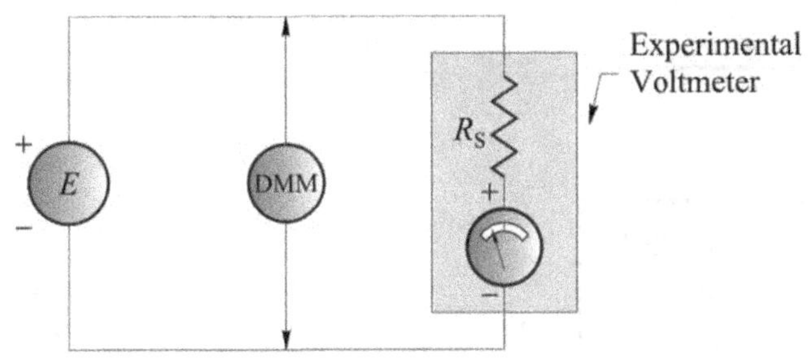

Figure 12-2

8. Increase the power supply voltage until a full-scale deflection is obtained on the experimental voltmeter. Measure the voltage for full-scale deflection with the DMM and record in Table 12-2.

9. Repeat steps 5 to 8 for ranges of 15 V and 10 V.

PART C: Loading Effect of a Voltmeter

10. Connect the circuit shown in Figure 12-3. Set the power supply voltage to 20 V (with a current limit of 10 mA if using a digital supply).

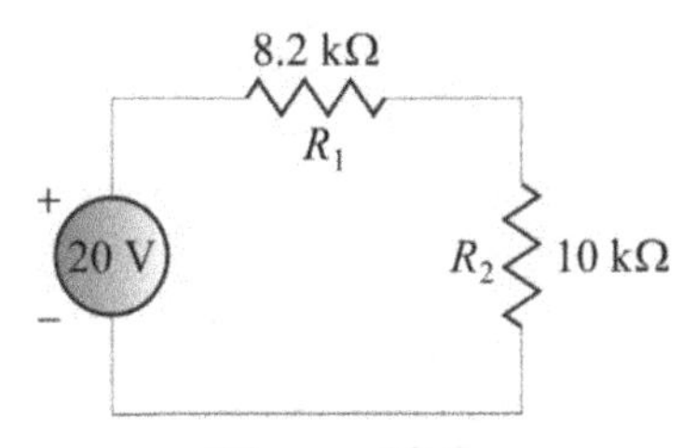

Figure 12-3

11. Using the 0-10 V experimental voltmeter from the previous procedure, measure the voltage across R_2. Record the measurement in Table 12-3.

12. Remove the experimental voltmeter and use the DMM to measure the voltage across R_2. Record the value in Table 12-3.

13. Calculate the voltage across R_2 and record in Table 12-3.

RESULTS

Table 12-1 Experimental Ammeter

Designed Range (mA)	Shunt Resistance (Ω)	Current for Full-Scale Deflection (mA)
2		
4		
11		

Table 12-2 Experimental Voltmeter

Designed Range (V)	Multiplier Resistance (Ω)	Voltage for Full-Scale Deflection (V)
5		
15		
10		

Table 12-3 Loading Effect of Voltmeter

V_{R2} Measured Exp. Voltmeter (V)	V_{R2} Measured DMM (V)	V_{R2} Calculated (V)

QUESTIONS

1. How do the measured values of meter ranges in Table 12-1 compare with design values? Explain any discrepancies.

2. Sketch a circuit for a multi-range ammeter that has the ranges shown in Table 12-1.

3. Referring to data from Table 12-2, how do the measured ranges compare with design values? Explain any discrepancies.

4. Calculate the % error caused by the voltmeter loading effect shown in Table 12-3. What voltmeter design considerations are important to minimize this effect?

5. A 100-mA, 1000-Ω movement is used to construct an ammeter. If the shunt resistor is 20.41 Ω, what is the range of this meter?

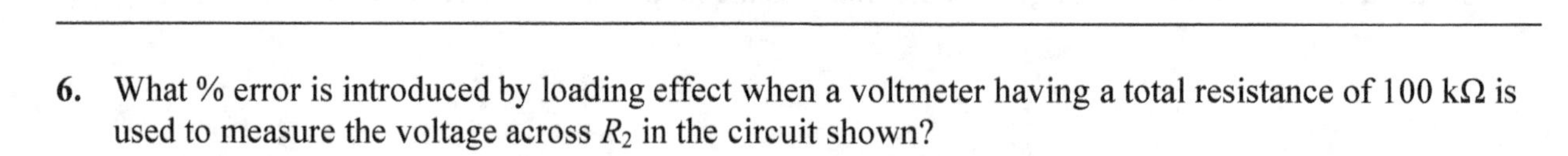

6. What % error is introduced by loading effect when a voltmeter having a total resistance of 100 kΩ is used to measure the voltage across R_2 in the circuit shown?

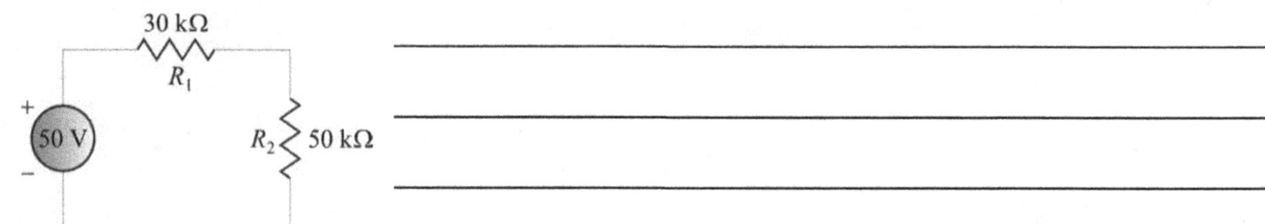

EXPERIMENT

13

CAPACITOR CHARGING AND DISCHARGING

OBJECTIVES

1. To determine the time constant of a capacitive circuit.

2. To investigate the exponential characteristic of a charging capacitor.

3. To investigate the exponential characteristic of a discharging capacitor.

DISCUSSION

Charging a Capacitor

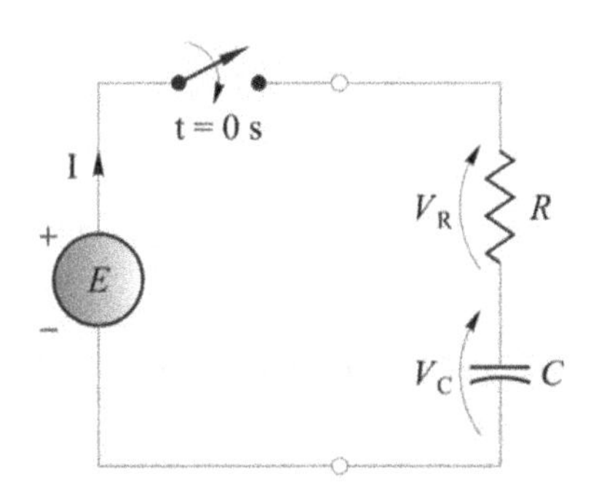

At the instant of closing the switch, there is no voltage across the capacitor, *C*. As current flows, a charge is built up on the capacitor and its voltage (V_C) increases until it eventually reaches source voltage (*E*).

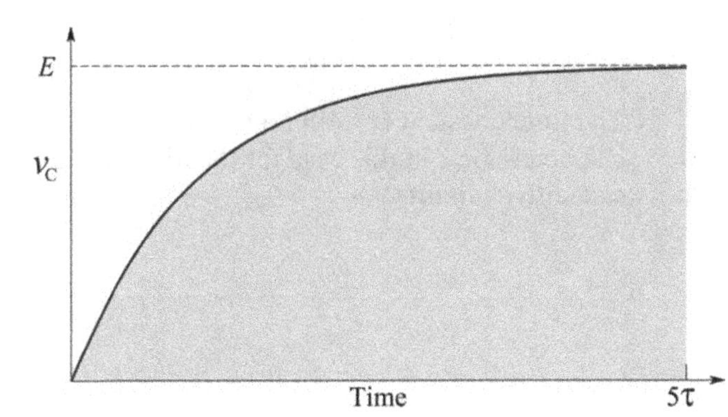

The rise in capacitor voltage is exponential and given by the following equation:

$$V_C = E\left(1 - e^{-t/RC}\right)$$

The product *RC* is defined as the time constant (τ) for the charging capacitive circuit.

Example: A circuit consisting of a 10-kΩ resistor in series with a 10-μF capacitor is suddenly switched across a 100-V DC source. Calculate:

(a) Time constant
(b) Voltage across the capacitor after 30 mS

(a) $\tau = RC = (10 \times 10^3) \times (10 \times 10^{-6}) = \mathbf{0.1\ s}$
(b) $V_C = E\left(1 - e^{-t/\tau}\right) = 100\left(1 - e^{-t/0.1}\right) = 100(1 - e^{-10t})$
When t = 30 ms,
$V_C = 100(1 - e^{-10\times0.03}) = 100(1 - e^{-0.3})$
$= 100(1 - 0.741) = \mathbf{25.9\ V}$

Discharging a Capacitor

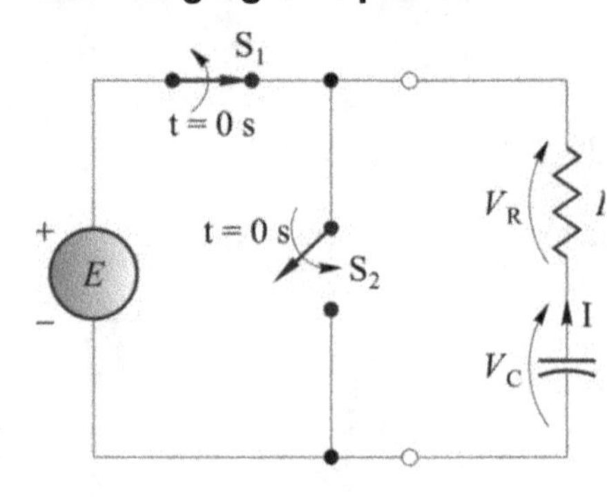

E has been applied for a long enough time (> 5τ) so that the capacitor has charged up to the source voltage (E). At t = 0 s, S_1 is opened and S_2 is closed, and the capacitor and resistor are shorted in series.

The capacitor now acts like a source and sends current in the opposite direction through the resistor. As it loses its charge, the capacitor voltage quickly falls, eventually reaching 0 V. The decay of voltage is again exponential.

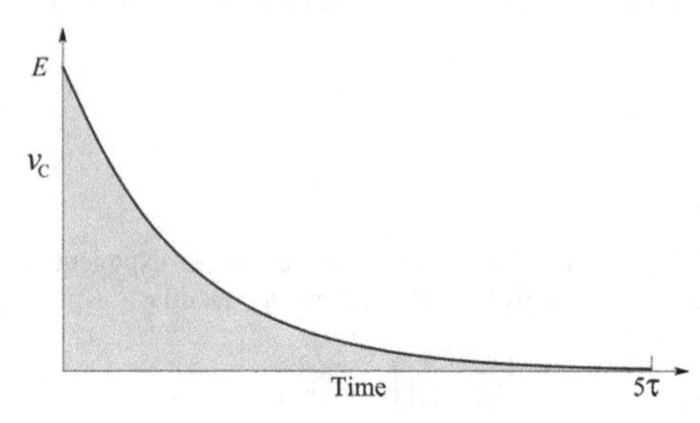

The fall in capacitor voltage is exponential and given by the following equation:

$$V_C = Ee^{-t/\tau}$$

The time constant (τ) still represents the RC characteristics, in this case for the discharging capacitive circuit.

EQUIPMENT

Power Supplies:
- Variable 0-24 DC, regulated

Instruments:
- Digital Multimeter (DMM)

Resistors: (5%, ½ W or higher)
- 1-kΩ
- Two 100-kΩ

Capacitors: *Electrolytic*
- 100-μF, 50-V DC

Miscellaneous:
- SPST switch
- SPDT switch
- A watch for timing (digital preferred)

PROCEDURE

PART A: Capacitor Charging

1. Connect the circuit shown in Figure 13-1. If the capacitor is polarized, make sure to observe correct polarity. Calculate the time constant of this circuit and record the result in Table 13-1.

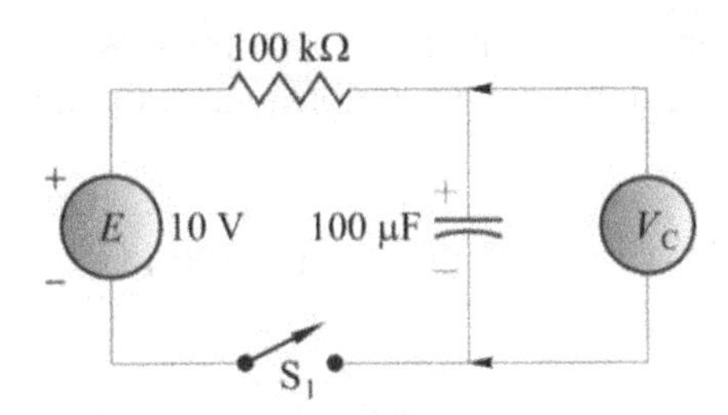

Figure 13-1

2. With switch S_1 **open**, adjust the source voltage E to 10 V (with a current limit of 10 mA if using a digital supply).

3. Connect the DMM across the capacitor as per the diagram. **Close** switch S_1 and measure the time taken for the voltage to reach 5 V. Record the value in Table 13-1. Calculate the time constant using the exponential formula and record the value in Table 13-1.

4. **Open** switch S_1 and connect the 1-kΩ resistor across the capacitor for 1 second to discharge it.

5. **Close** switch S_1 and measure capacitor voltage at the time intervals shown in Table 13-1. Calculate the theoretical values of V_C using the exponential formula and record them in Table 13-1.

6. Connect the circuit shown in Figure 13-2. Calculate the time constant of this circuit and record the value in Table 13-2.

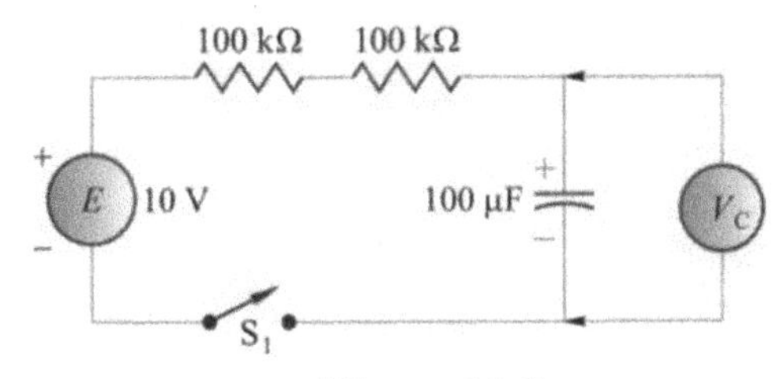

Figure 13-2

7. Repeat steps 3 to 5 and record results in Table 13-2.

PART B: Capacitor Discharging

8. Connect the circuit shown in Figure 13-3.

9. **Close** the switch to position 1 and adjust the power supply to 10 V (with a current limit of 15 mA if using a digital supply). The capacitor will charge rapidly through the 1-kΩ resistor.

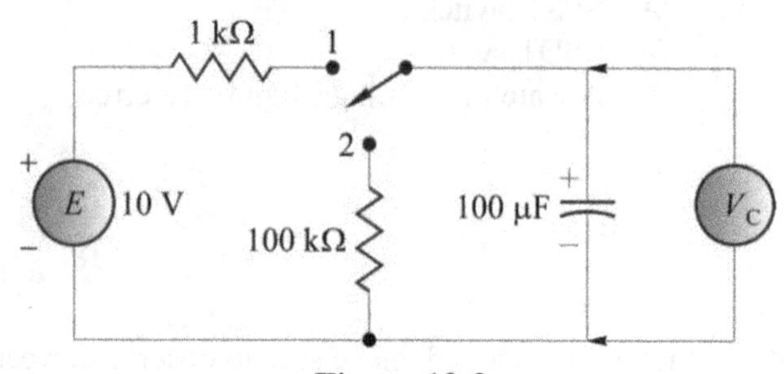

Figure 13-3

10. Move the switch to position 2 and measure the capacitor voltage as it discharges at the time intervals shown in Table 13-3. Calculate the theoretical values of V_C using the exponential formula and record them in Table 13-3.

11. Turn **off** the power supply. Connect a second 100-kΩ resistor in series with the existing 100-kΩ resistor and repeat steps 9 and 10.

12. On the same set of axes, plot graphs of measured V_C versus t for each resistance for the charging and discharging circuits.

RESULTS

Table 13-1 Capacitor Charging (R = 100 kΩ)

Resistance		τ (Calc. RC)		Time for 5 V		τ	
100 kΩ							
t (s)	10	20	30	40	50	60	80	100
V_C (meas.)								
V_C (calc.)								

Table 13-2 Capacitor Charging (R = 200 kΩ)

Resistance		τ (Calc. RC)		Time for 5 V		τ	
200 kΩ							
t (s)	10	20	30	40	50	60	80	100
V_C (meas.)								
V_C (calc.)								

Table 13-3 Capacitor Discharging

R	t (s)	10	20	30	40	50	60	80	100
100 kΩ	V_C (meas.)								
	V_C (calc.)								
200 kΩ	V_C (meas.)								
	V_C (calc.)								

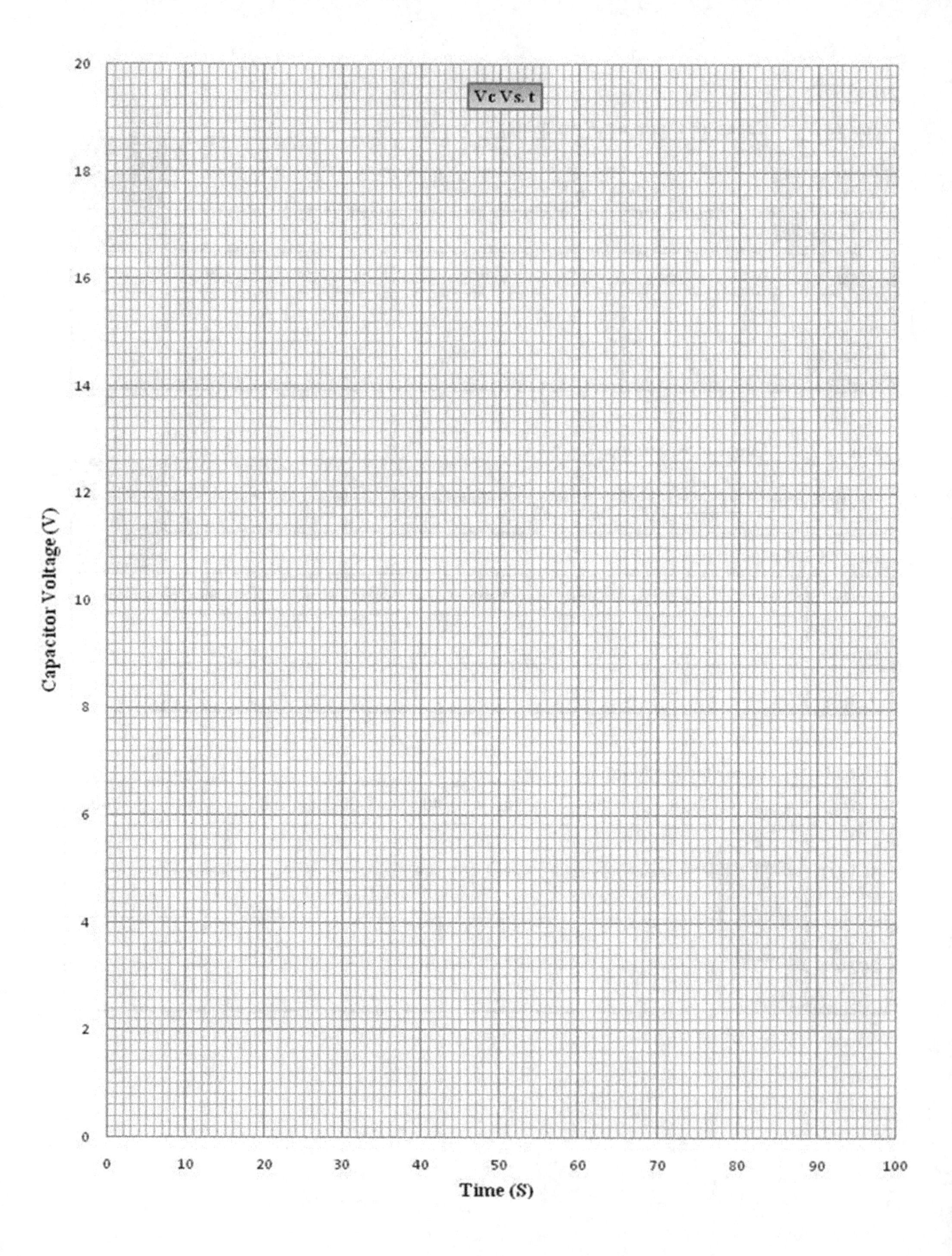
Vc Vs. t
Capacitor Voltage (V)
Time (S)
0
2
4
6
8
10
12
14
16
18
20
0
10
20
30
40
50
60
70
80
90
100

QUESTIONS

1. How do the measured values of time constants in Tables 13-1 and 13-2 compare to calculated values? Explain any discrepancies.

2. Comment on the shape of your plotted graphs. Do they conform to the theoretical curves?

3. What error is introduced by the connection of the voltmeter in the charging and discharging circuit?

4. A 10-μF capacitor is charged up to 500 V and left on a laboratory bench. A student named G.I. Hert accidentally picks up the capacitor by the leads and after 2 seconds the voltage has fallen to 300 V. Calculate the resistance of G.I. Hert's body. *(Warning: Do not recreate this scenario in a lab setting!)*

5. Why is it important to presume capacitors are fully charged before handling them? Explain your answer. Provide some examples of how to handle capacitors with extreme care.

EXPERIMENT

14

OSCILLOSCOPE AND FUNCTON GENERATOR OPERATION

OBJECTIVES

1. To identify and adjust the operating controls of a basic digital oscilloscope.
2. To observe the internal calibration voltage on the oscilloscope.
3. To identify the features and controls of a function generator.
4. To operate the function generator and observe output waveforms on the oscilloscope.

DISCUSSION

Oscilloscope

The digital oscilloscope, also known as a **scope**, is a complex electronic device that combines software and hardware to acquire, process, and display signals. In some models, they can even store the captured signals._The digital oscilloscope is one of the basic instruments for measuring AC quantities. It has the ability not only to *measure* AC quantities such as frequency and voltage but also to *display* the waveform from which these quantities are derived.

The oscilloscope draws a graph on a tin display panel or LCD screen and a grid of lines is etched on the faceplate to serve as a reference for measurements. This is referred to as the **graticule**.

Most lab oscilloscopes are **dual-trace** (2- channel) scopes, but four channel scopes are also available. The two traces are developed on the screen by electronic switching. This makes it possible to observe simultaneously two time varying waveforms.

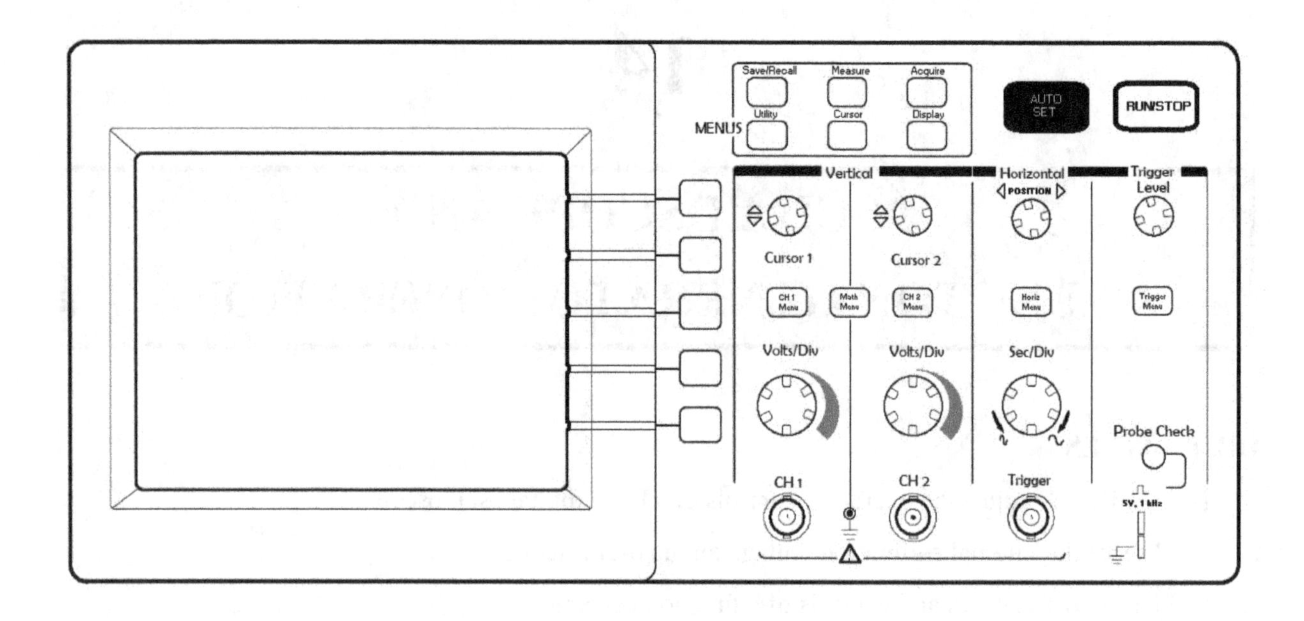

Vertical Controls

CH1, CH2, Cursor 1, and Cursor 2: Position the waveforms vertically on the screen.

CH1 and CH2 Menu: Display the vertical menu selections and toggle the display of the channel waveform ON and OFF.

Volts/Division: This sets the vertical sensitivity of each channel. The **graticule** is divided up into centimeter squares and each centimeter will represent a certain voltage depending on the volts/div setting. This is used to adjust the height of the waveform to obtain an accurate measurement. There is also a **variable** volts/division incorporated into this control to allow for a fine adjustment of the vertical height. Normally this control is in the **cal** position (fully clockwise) which means the control is now calibrated to the volts/div specified on the main control.

MATH Menu: Displays all math operations available with the signals of CH1 and CH2. Also toggles the display of the math waveform ON and OFF.

Horizontal Controls

Position: Adjusts the horizontal position of CH1, CH2, and MATH waveforms on the screen.

HORIZ Menu: Displays the horizontal menu.

Sec/Div: Allows the selection of the horizontal scale factor (time/div) to display the waveforms on the screen.

Trigger Controls

Trigger Level: Sets the amplitude level the signal must cross to be acquired.

TRIG Menu: Displays the Trigger Menu

Menu and Control Buttons

Save/Recall: Displays the Save/Recall Menu for all setups and waveforms.

Measure: Displays the automated measurement menu.

Acquire: Displays the modes to acquire waveforms menu.

Display: Displays the display modes menu.

Cursor: Displays all cursor menus. The vertical position controls are displayed to allow the measurement of time differences. The horizontal position controls are displayed to allow the measurement of voltage differentials.

Function Generator

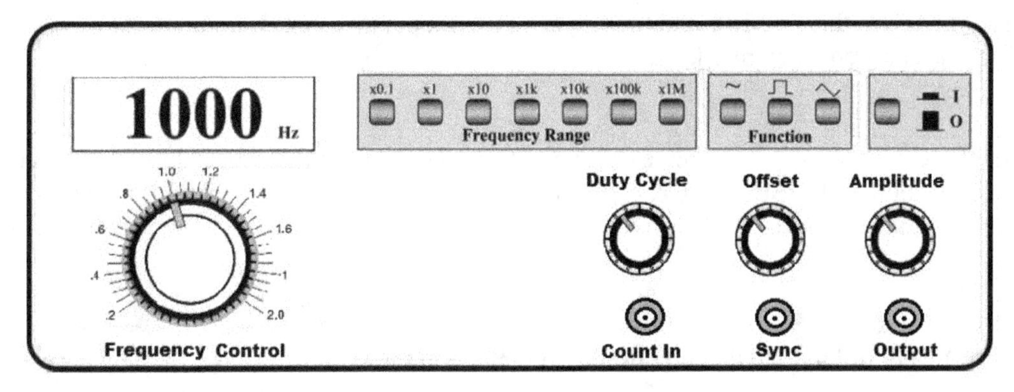

A function generator produces variable frequency AC voltage waves. The waveforms can be sinusoidal, square, or triangular with frequencies from a few hertz to the megahertz range. A function generator is used for testing and experimentation where different AC waves at varying frequencies are required.

Operating Controls

Waveform selector: Selects either sine, square, or sawtooth waveform. Sometimes called the **function switch**.

Output level: Controls the amplitude of the output voltage.

Range Switch: Usually a series of push buttons to select the output frequency range.

Frequency Control: Provides for an exact setting of frequency within the selected range with a rotary control. Often *coarse* and *fine* controls are provided.

Output Jack: Where the waveform selected by the function switch is available.

EQUIPMENT

Power Supplies:

- Function generator

Instruments:

- Oscilloscope

PROCEDURE

PART A: Oscilloscope Operation

1. With the power to the oscilloscope **off**, examine the front panel noting the types and functions of each control and jack.

2. Power on the oscilloscope and connect the probe to CH1. To do so, align the slot of the probe to the CH1 BNC connector, push, and twist to the right to lock the scope probe in place. Connect the probe tip to the PROBE CHECK connectors.

3. Push the AUTO SET button. After a few seconds, you should see a square wave present on the screen of the oscilloscope. Adjust the Volt/Div and time base on the scope; this signal should be approximately 5 V_{PP} at a frequency of 1 kHz.

4. Push the CH1 MENU button twice to remove channel 1 from the display, and proceed to disconnect the CH1 probe tip from the PROBE CHECK connector. Now connect the second oscilloscope probe between the CH2 and the PROBE CHECK connector to verify its operation. Sketch the waveform in Figure 14-2a.

5. Change the Volt/Div control to 2V/Div and observe the effect on the waveform. Sketch the waveform in Figure 14-2b.

6. Change the time base Sec/Div control to 0.5 ms/div and observe the effect on the waveform. Sketch the waveform in Figure 14-2c.

PART B: Function Generator Operation

7. Examine your function generator, noting the controls and output connections.

8. Connect the output of the function generator to the CH1 input of the oscilloscope as shown in Figure 14-1. Turn on the function generator. Set the function generator to sine (~) output with a frequency of 1000 Hz. Adjust the output voltage level to the middle of its range.

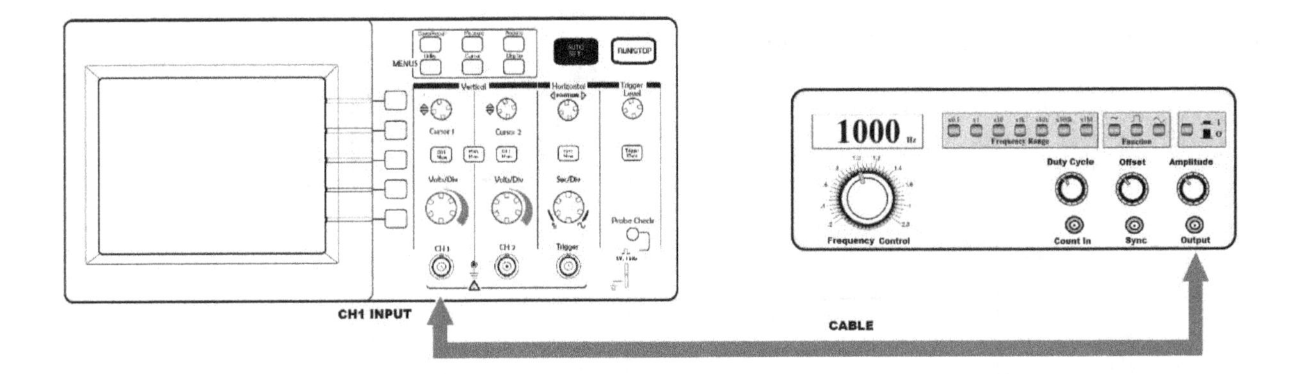

Figure 14-1

9. Turn **on** the oscilloscope and adjust the sec (time)/div control to display one cycle across the width of the screen. Adjust the volts/div control for a waveform height of approximately six divisions from peak-to-peak. Sketch the waveform displayed in Figure 14-3a

10. Change the frequency to 2000 Hz. **Do not** change the sec (time)/div control. Centre the display vertically and horizontally and readjust the height to six divisions if necessary. Sketch the waveform displayed in Figure 14-3b.

11. Change back to 1000 Hz frequency and set the function switch to produce a **triangular wave** output. **Do not** change the sec (time)/div control. Centre the display vertically and horizontally and readjust the height to six divisions, if necessary. Sketch the waveform displayed in - Figure 14-4a.

12. Repeat step 11 for the triangular wave. Sketch the waveform displayed in Figure 14-4b.

13. Change back to 1000 Hz frequency and set the function switch to produce a **square wave** output. **Do not** change the sec (time)/div control. Center the display vertically and horizontally and readjust the height to six divisions, if necessary. Sketch the waveform displayed in Figure 14-5a.

14. Repeat step 11 for the square wave. Sketch the waveform displayed in Figure 14-5b.

RESULTS

Figure 14-2 Oscilloscope Operation

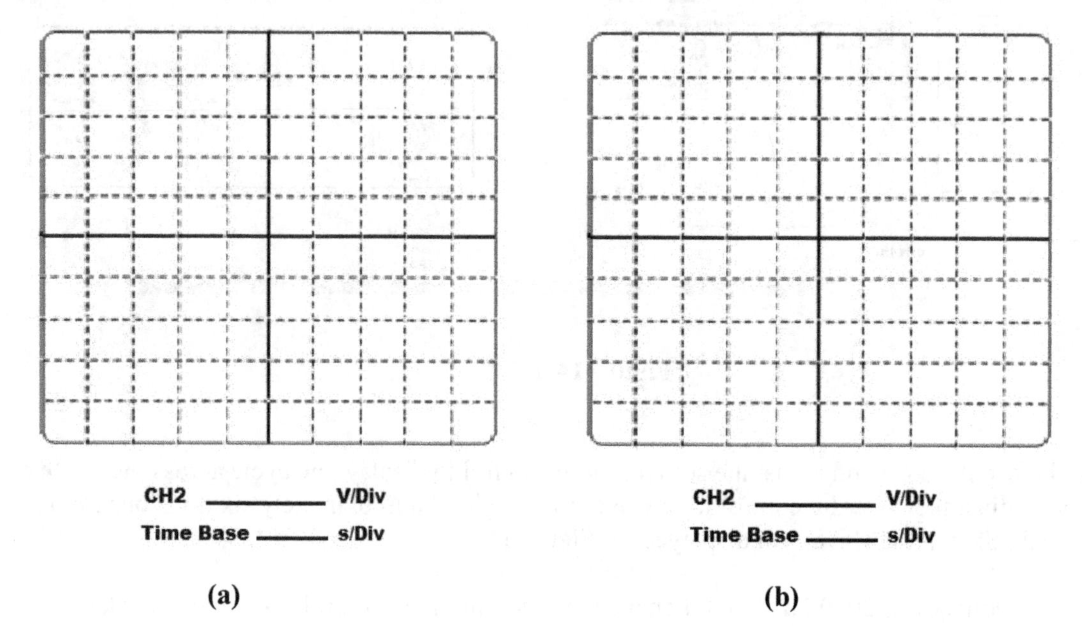

(a) **(b)**

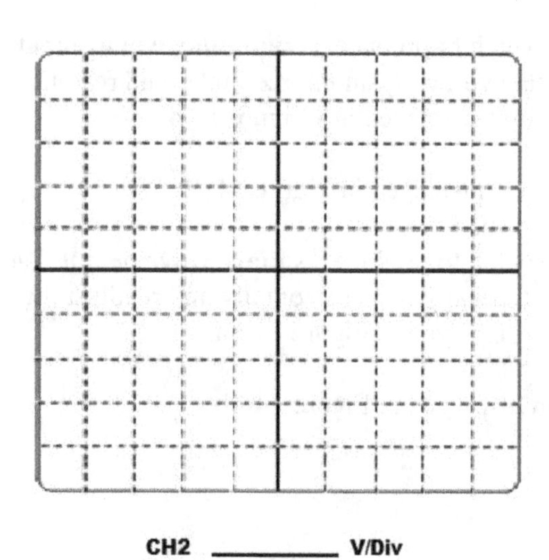

(c)

Figure 14-3 Sine Wave

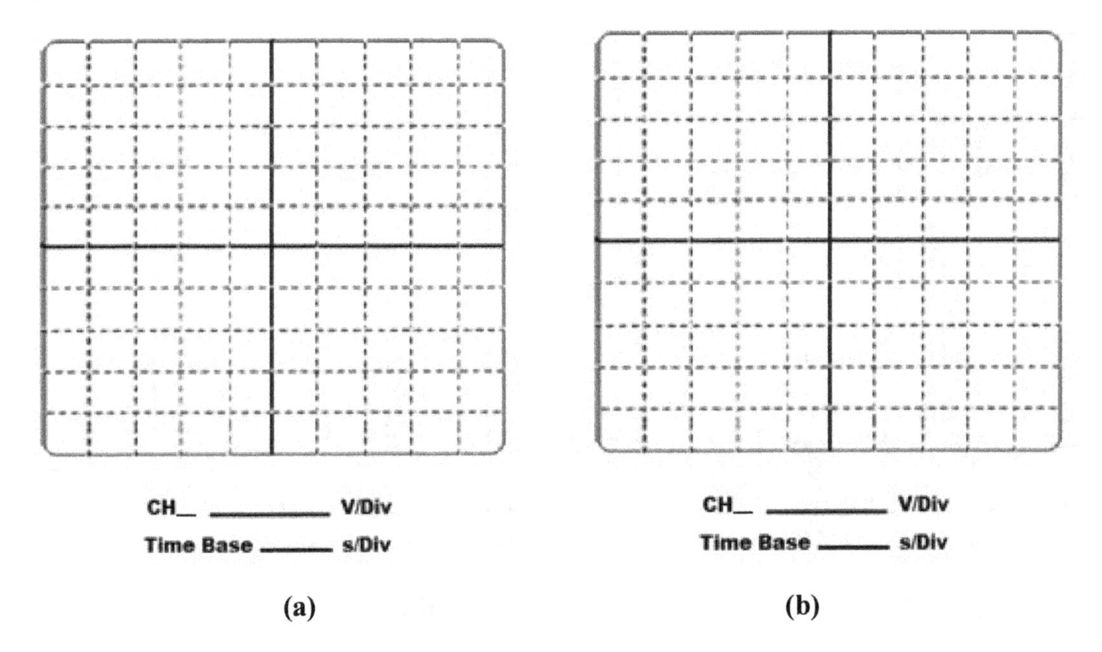

Figure 14-4 Triangular Wave

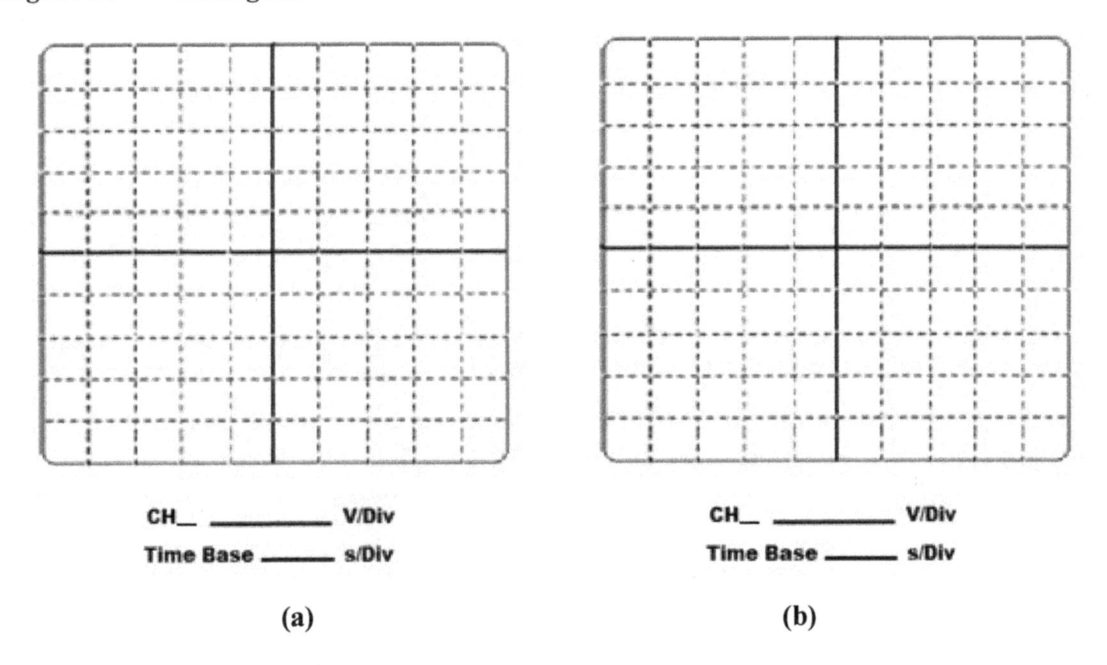

Figure 14-5 Square Wave

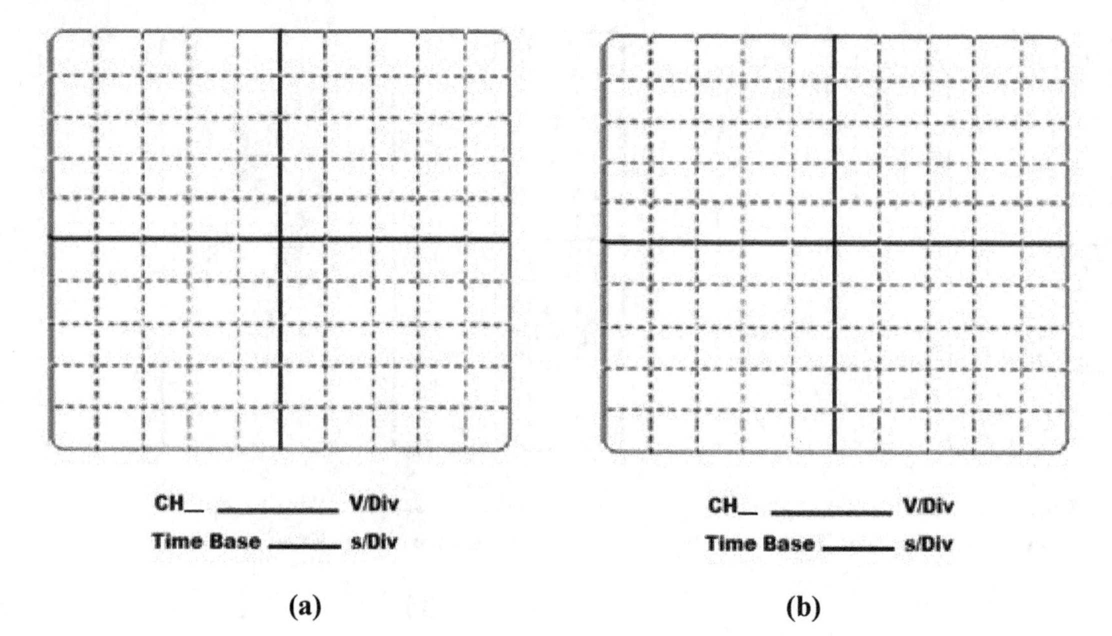

QUESTIONS

1. In procedure step 6, the sec (time)/div control was changed from its original setting. Describe the effect this had on the number of cycles displayed.

____

____

____

____

2. Refer to Question 1. Was the frequency of the calibration waveform changed when changes were made to the sec (time)/div setting in step 6? Explain your answer.

____

____

____

____

3. Referring to your oscilloscope, which controls affect the following?

(a) The height of the displayed waveform ______________________________

(b) The number of cycles displayed ______________________________

(c) The starting point of the waveform ______________________________

(d) The position of the waveform ______________________________

(e) Engaging the sweep generator ______________________________

4. In your own words, discuss the relationship between the number of cycles displayed on the screen and the frequency setting of the function generator (with a constant sec (time)/div setting). Refer to your data.

____

____

____

____

____

5. A function generator on sine output is adjusted to 5-V peak voltage at a frequency of 10-kHz. It is connected to an oscilloscope with a sec (time)/div setting of 0.2 ms and a volts/div setting of 2 V. Sketch on the grid below the waveform the scope will display.

6. Use Multisim and connect the function generator on sine wave and set the output voltage to 5-V peak and a frequency of 10-kHz. Connect the function generator to an oscilloscope with a sec (time)/div setting of 0.2 ms and a volts/div setting of 2 V. Compare the waveform displayed on the oscilloscope with the one you sketched in question 5.

____

____

____

____

____

EXPERIMENT

15

OSCILLOSCOPE VOLTAGE AND FREQUENCY MEASUREMENTS

OBJECTIVES

1. To measure peak-to-peak voltages with the oscilloscope.
2. To measure peak-to-peak voltages using the cursors
3. To measure DC voltages with the oscilloscope.

DISCUSSION

Voltage Measurement

AC voltages are measured on the oscilloscope by measuring the height of the waveform and using this in conjunction with the volts/div setting to determine the peak or peak-to-peak voltage.

Example: For the display shown, if the volts/div setting is 5 V/div, determine the peak voltage.

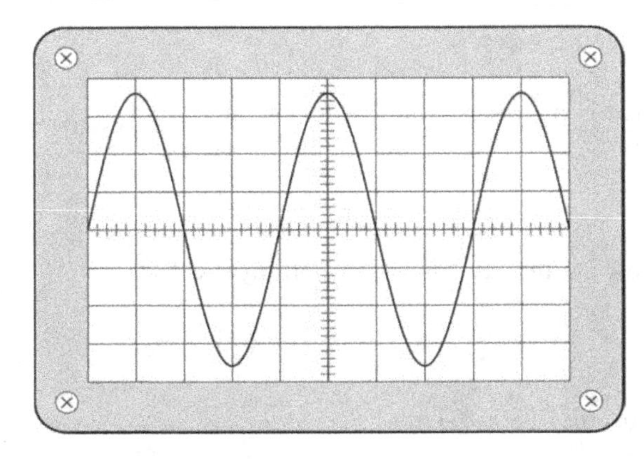

$$V_{\text{peak}} = 3.6\ \text{div} \times 5\ \text{V/div} = \mathbf{18\ V}$$

Note: The coupling of the oscilloscope channel should be set to the AC setting to remove any DC level.

DC voltages are represented by a vertical displacement of the trace above or below the reference axis.

Example: For the display shown, if the volts/div setting is 2 V/div, determine the DC voltage.

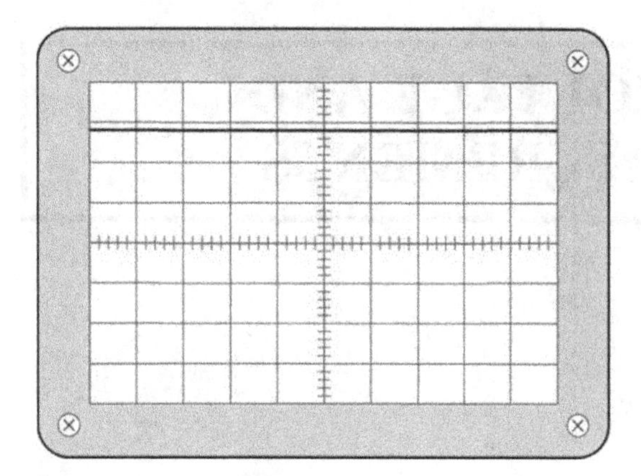

$$V_{dc} = 2.8 \text{ div} \times 2\text{ V/div} = \mathbf{5.6\text{ V}}$$

Note: The coupling on the oscilloscope channel should be set to the DC setting.

Oscilloscope Self-Calibration

The self-calibration routine of the oscilloscope is performed to optimize the reading measurements of the instrument. Traditionally, this routine should be performed if the temperature in the room where the oscilloscope is located varies by 5° C or more. To do so, check the manual of the instrument and follow the instructions to perform the self-calibration.

Frequency Measurement

The oscilloscope can also measure the frequency of a periodic wave.

The number of divisions occupying one cycle of the wave is measured and this is converted to a cycle time by multiplying by the sec (time)/div setting. $f = \frac{1}{T}$

Example: Determine the frequency of the waveform shown if the sec (time)/div setting is 0.1 ms.

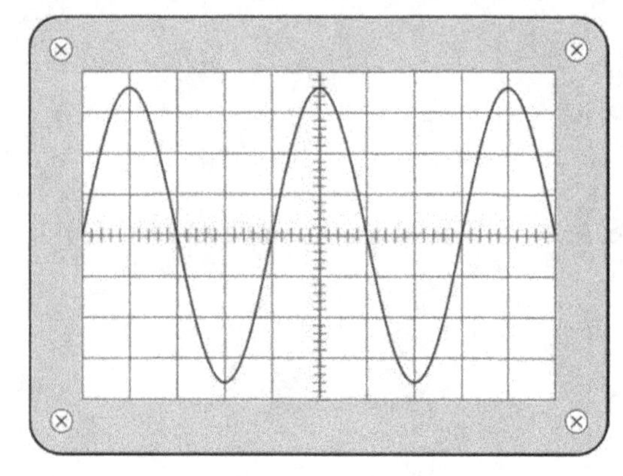

$$T = 4 \text{ div} \times 0.1\text{ ms/div} = 0.4\text{ ms}$$

$$f = \frac{1}{T} = \frac{1}{0.4\text{ ms}} = \mathbf{2.5\text{ kHz}}$$

Digital oscilloscopes can take automatic measurements, such as frequency, period, peak-to-peak voltages, rise time, fall time, and, in some instances, phase-shift and duty cycle. To make these types of measurements, follow the steps listed below:

1. Press the MEASURE button to see the Measure Menu displayed on the screen.

2. Push the top option button. This will show the Measure 1 Menu. Continue pressing to select the type of measurement desired. If you select Freq., this is what will be displayed:

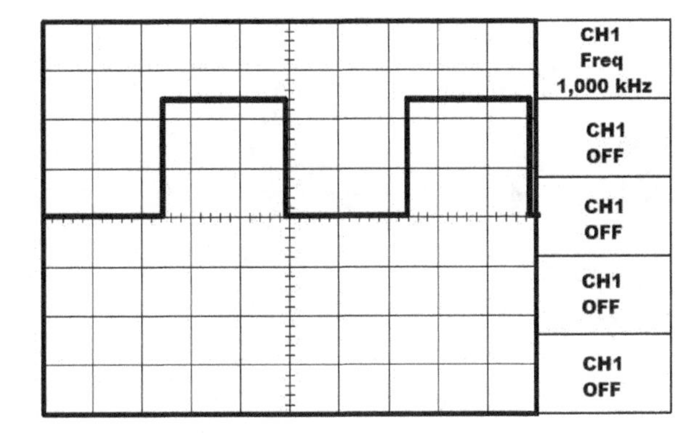

3. Push the Back option button

4. Push the second option button from the top. The Measure 2 Menu will appear. Continue pressing to select the type Pk-Pk:

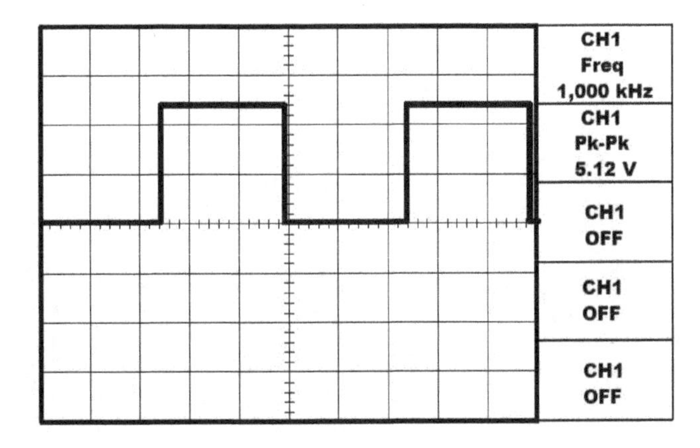

Cursors

Another method of measurement using a digital oscilloscope is by means of the cursors. These will appear in pairs, read the numeric value of their placement, and display these values on the screen. Two types of cursors are available on the oscilloscope: the Voltage and Time cursors.

When using the cursors, select the source of the waveform on the display that you want to measure.

To use the cursors, press the CURSOR button and select Voltage or Time. If you select Voltage, the cursors will appear as horizontal lines on the display and will measure the vertical parameters on the screen (Voltage). If Time is selected, vertical lines will appear and read the horizontal parameters on the screen (time).

When the cursors (Voltage or Time) are selected, the difference between them will appear as a Delta, as shown below:

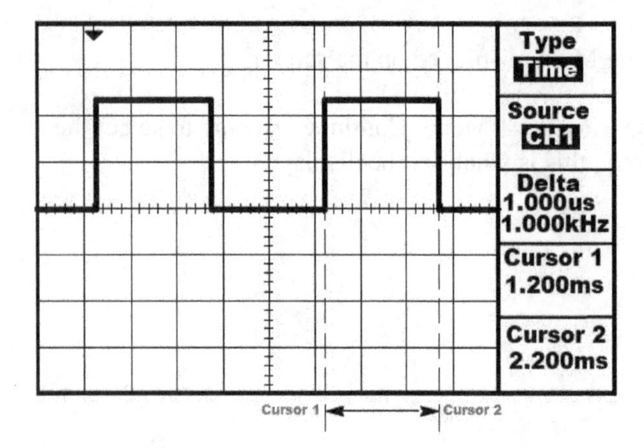

EQUIPMENT

Power Supplies:
- Isolation transformerDC power supply
- Function generator

Instruments:
- Oscilloscope
- DMM

Resistors: (5%, ½ W or higher)
- One each of 4.7 kΩ, 8.2-kΩ, and15-kΩ

Other:
- Protoboard
- Hook-up Wire

PROCEDURE

PART A: AC Voltage Measurement

1. Connect the circuit as shown in Figure 15-1.

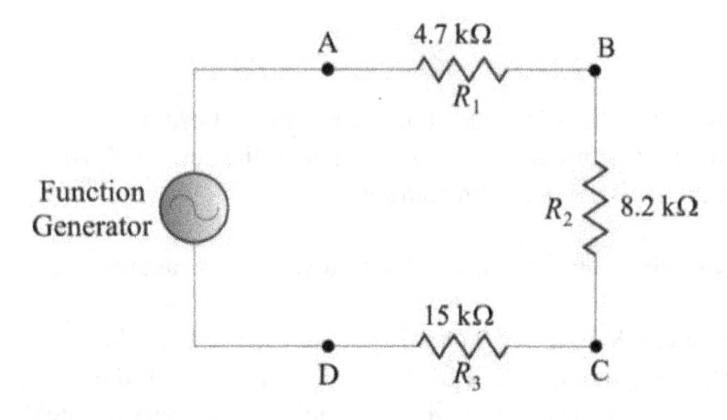

Figure 15-1

2. Turn on the function generator. Set the amplitude to the lowest value and the frequency to 60 Hz.

3. Increase the output of the function generator until the voltage across points AD is 20V peak-to-peak. *Note*: The DMM measures RMS voltages, so you will have to calculate the equivalent RMS voltage.

4. When making measurements with an oscilloscope, it is important to centre the trace on the horizontal axis. Use the following procedure before making any measurements. Press the CH1 Menu. Under coupling, switch to the GND setting. This will ground the channel 1 and allow you to adjust the controls to produce a trace centered on the horizontal axis. Then, under the same CH1 Menu and coupling, switch back to the AC or DC position depending on the measurement to be performed.

5. Connect the scope probe across terminals AB and adjust the controls to obtain several cycles of voltage. Adjust the volts/div control to obtain the largest measurable display. Using the Measured Menu, record the peak-to-peak voltage value of the signal in Table 15-1.

6. Repeat step 5 with the probe connected across BC, CD, and BD.

7. Calculate, using circuit analysis, the peak-to-peak voltages across AB, BC, CD, and BD, assuming a source voltage of 20 V peak-to-peak. Record your answers in Table 15-1.

PART B: DC Voltage Measurement

8. Connect the series circuit used in Part A across a DC power supply as shown in Figure 15-2.

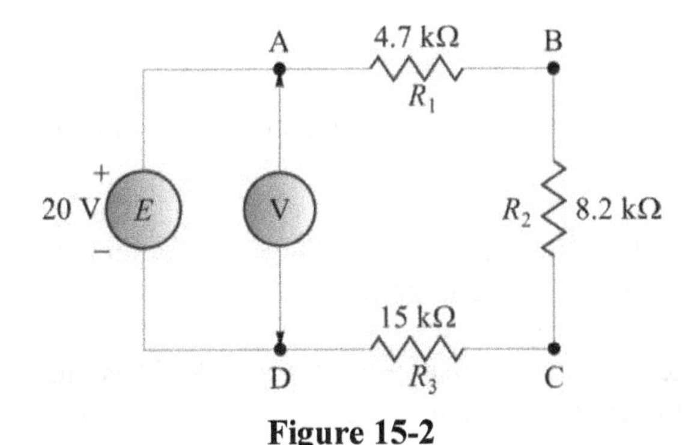

Figure 15-2

9. Turn the power supply **on** and adjust the voltage across terminals AD to 20 V.

10. Using the oscilloscope, press the CH1 Menu. Under coupling, switch to the GND setting. Use the vertical controls to adjust the reference for DC voltage measurements. Connect the scope across points AB and switch to DC coupling using the CH1 menu under coupling.

11. Using the Measured Menu, record the peak-to-peak voltage value of the signal in Table 15-2.

12. Repeat step 11 with the probe connected across BC, CD, and BD.

13. Using the CURSORS in Voltage mode, record the peak-to-peak voltage of the input signal in Table 15-2.

14. Repeat step 13 with the probe connected across BC, CD, and BD.

15. Measure voltages V_{AB}, V_{BC}, V_{CD}, V_{BD} with the DMM and record the values in Table 15-2.

16. Calculate voltages V_{AB}, V_{BC}, V_{CD}, V_{BD} from circuit analysis using a supply voltage of 20 V. Record the values in Table 15-2.

PART C: Frequency Measurement

17. Connect the circuit as shown in Figure 15-3. The output of the function generator is connected to CH1 or A input.

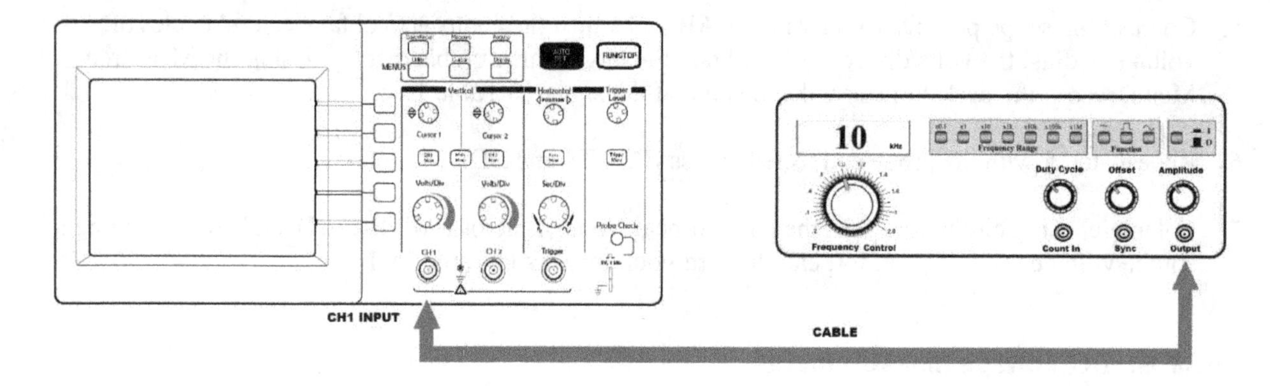

Figure 15-3

18. Turn **on** the function generator and adjust it for sine wave output with amplitude set approximately midway. Adjust the frequency control to 10 kHz.

19. With the oscilloscope's coupling in the GND mode, adjust the controls to produce a horizontal trace centred vertically. Switch the coupling under the CH1 Menu to AC coupling.

20. Adjust the sec (time)/div control to display two cycles or more and using the MEASURED Menu, measure the frequency of the signal. Record this value control in Table 15-3.

21. Using the CURSORS in Time mode, record the frequency voltage of the signal in Table 15-3.

22. Adjust the frequency of the function generator to 50 kHz and repeat steps 19, 20, and 21.

23. Adjust the frequency of the function generator to 500 Hz and repeat steps 19, 20, and 21.

RESULTS

Table 15-1 AC Voltage Measurement

Test Points	Voltage	MEASURED MENU **Measured Voltage (V)**	USING CURSORS **Measured Voltage (V)**	CIRCUIT **Calculated Voltage (V)**
A to B	V_{AB}			
B to C	V_{BC}			
C to D	V_{CD}			
B to D	V_{BD}			

Table 15-2 DC Voltage Measurements

Test Points	Voltage	SCOPE **Measured Voltage (V)**	DMM **Calculated Voltage (V)**	CIRCUIT **Calculated Voltage (V)**
A to B	V_{AB}			
B to C	V_{BC}			
C to D	V_{CD}			
B to D	V_{BD}			

Table 15-3 Frequency Measurement

Generator Frequency	MEASURED MENU **Measured Frequency (kHz)**	USING CURSORS **Measured Frequency (kHz)**
10 kHz		
50 kHz		
500 Hz		

QUESTIONS

1. In your own words, explain how AC voltages are measured with an oscilloscope.

____

____

____

____

2. Compare the voltage measurements made with the oscilloscope (Measured Menu and Cursors) to the calculated circuit values for Part A. Refer to your data and discuss any discrepancies.

____

____

____

____

3. What is the oscilloscope's main advantage for making AC voltage measurements?

____

____

____

____

4. What is the oscilloscope's major disadvantage for making AC voltage measurements?

____

____

____

____

5. Compare the frequencies measured with the oscilloscope (Measured Menu and Cursors) to the function generator setting. What are the main sources of error in this measurement?

6. An oscilloscope displays the waveform shown below. If the volts/div setting is set on 10 and the sec/div setting is on 1.0 ms, determine the peak-to-peak voltage and the frequency.

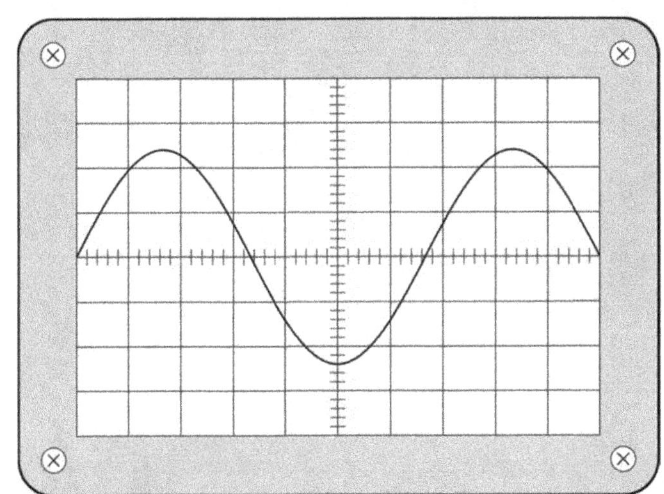

EXPERIMENT

16

PEAK AND RMS AC VOLTAGES

OBJECTIVES

1. To investigate the relationship between peak and RMS values of AC voltages and currents.
2. To verify this relationship by experimentation.

DISCUSSION

AC Voltage Wave

An AC voltage wave is a sine wave that alternates from positive to negative at a specified frequency. For example a 60-Hz AC voltage will make 60 complete alternations in 1 second period.

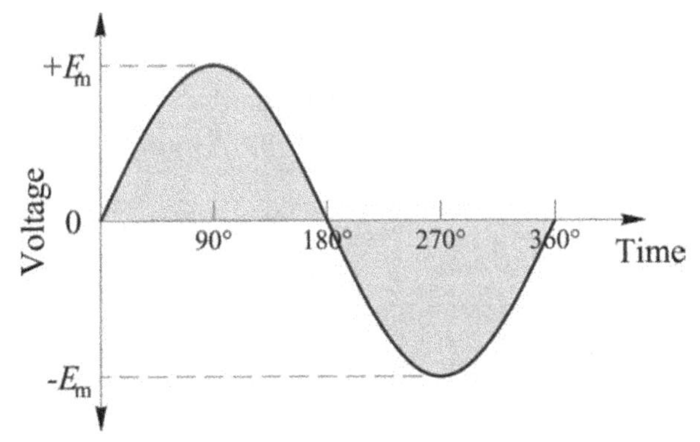

The equation for this voltage is

$$e = E_m \sin 2\pi f t$$

The maximum or *peak* value is E_m.

RMS Value of Voltage

Since the AC voltage varies continuously from maximum positive to maximum negative value, it does not have the same effect or produce the same power as a steady DC voltage with the same voltage as the peak AC voltage. The **effective** *or* **RMS** (root mean square) value of the AC voltage is the value that will produce the same power as the equivalent DC voltage. From power relationships it can be shown that the effective *or* RMS value of the AC voltage is given by the following equation:

$$E_{\text{RMS}} = 0.707E_{\text{m}}$$

This value is the voltage that an AC meter will measure and is normally used in calculations involving AC circuits. An oscilloscope displays the AC wave and can be used to measure the *peak* value of the AC voltage.

Example: A voltage wave is given by the equation $e = 163\sin 377t$. If the source is connected to a 100 Ω resistor, calculate:

(a) RMS value of voltage.
(b) RMS current in the resistor.

(a) $E_{\text{m}} = 163\text{ V} \Rightarrow E_{\text{RMS}} = 0.707 \times 163\text{ V} = \mathbf{115.2\ V}$

(b) $I_{\text{RMS}} = \dfrac{E_{\text{RMS}}}{R} = \dfrac{115\text{ V}}{100\,\Omega} = \mathbf{1.15\ A}$

EQUIPMENT

Power Supplies:
- Function generator

Instruments:
- Oscilloscope
- DMM

Other:
- Protoboard
- Hook-up Wire

Resistors: (5%, ½ W or higher)
- One each of 100-Ω, 220-Ω, 330-Ω, 820-Ω, 1-kΩ, and 1.5-kΩ

PROCEDURE

PART A: Line Voltage Supply (60 Hz)

1. Connect the circuit shown in Figure 16-1. Set the function generator to its lowest voltage and a frequency of 60 Hz. Switch it **on** and adjust the output of the function generator to 5 V_{RMS} and keep it at this level for the remainder of Part A. Record this voltage (V_{AD}), in Table 16-1.

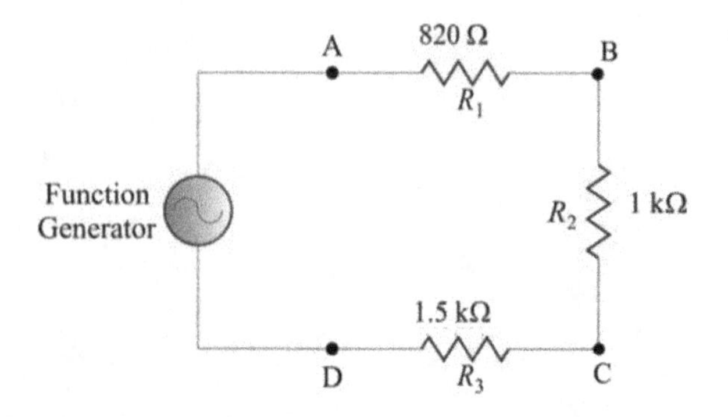

Figure 16-1

2. Measure the voltage across each resistor with the DMM and record the values in Table 16-1 under V_{rms}, *measured.* Turn off the function generator.

3. Connect the oscilloscope across the output of the function generator points AD). Turn it **on** and measure the peak voltage using the Measure Menu. Record this value in Table 16-1 under V_P, *measured.*

4. Using the oscilloscope, measure the peak voltage across each resistor and record the values in Table 16-1 under V_P, *measured* using the Measure Menu. Turn **off** the function generator.

5. Connect a DMM in the current mode to measure the RMS current flowing in the circuit. Turn **on** the function generator and measure the current. Record the value in Table 16-1 under I_{rms} *measured.*

6. Using circuit analysis, calculate the RMS circuit current and RMS and peak voltages across each resistor and record in Table 16-1.

PART B: Function Generator Supply (2000 Hz)

7. Connect the circuit in Figure 16-2.

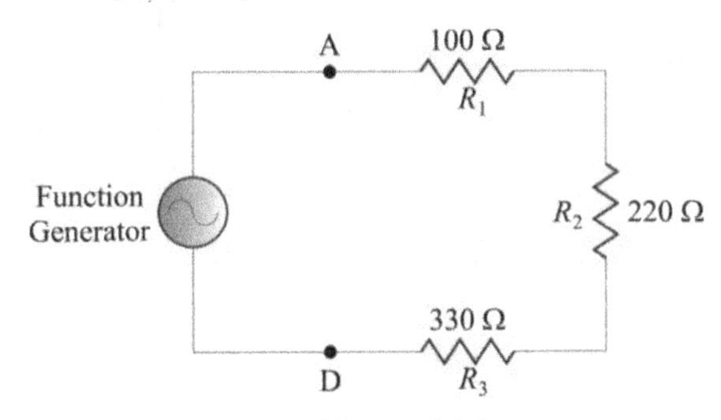

Figure 16-2

8. Turn **on** the function generator and set the frequency to 2000 Hz. Using the DMM, adjust the output voltage of the function generator to 5V $_{RMS}$.

9. Using the DMM, measure the voltage across each resistor and record in Table 16-2 under V_{rms}, *measured.*

10. Using the oscilloscope, measure the peak voltage across the generator output (V_{AD}) and across each resistor using the MEASURE Menu. Record these values in Table 16-2 in V_P, *measured* column.

11. Switch off the function generator and connect the AC ammeter to measure the RMS current flowing in the circuit. Switch on the function generator, readjust the voltage to 5 V_{RMS} and frequency to 2000 Hz if necessary, and measure the circuit current. Record value in Table 16-2 under I_{rms}, *measured.*

12. Using circuit analysis, calculate the RMS circuit current and RMS and peak voltages across each resistor and record in Table 16-2.

RESULTS

Table 16-1 Function Generator at 60 Hz

	V_{RMS} (V)		V_{PK} (V)		I_{RMS} (mA)	
	Meas.	Calc.	Meas.	Calc.	Meas.	Calc.
V_{AD} (V)	40 V					
R_1						
R_2						
R_3						

Table 16-2 Function Generator at 2000 Hz

	V_{RMS} (V)		V_{PK} (V)		I_{RMS} (mA)	
	Meas.	Calc.	Meas.	Calc.	Meas.	Calc.
V_{AD} (V)	10 V					
R_1						
R_2						
R_3						

QUESTIONS

1. Compare measured peak voltages to calculated voltages in Table 16-1. Explain any discrepancies.

____

____

____

____

____

2. Calculate the ratio V_{rms}/V_P for each measured value of voltage in Table 16-1. Compare your values to the theoretical constant of 0.707. Explain any discrepancies.

____

____

____

____

____

3. Does your current data verify that Ohm's law is applicable to AC circuits? Refer to your results.

____

____

____

____

____

____

4. A voltage source has a waveform given by the following equation $e = 150\, sin\, 377t$. What is the RMS voltage of this source? What will an AC ammeter indicate if the source is connected to a 10-Ω resistor?

EXPERIMENT

17

CAPACITIVE REACTANCE

OBJECTIVES

1. To investigate the effect of capacitance and frequency on capacitive reactance.
2. To investigate the phase shift caused by a pure capacitance.

DISCUSSION

Capacitive Reactance

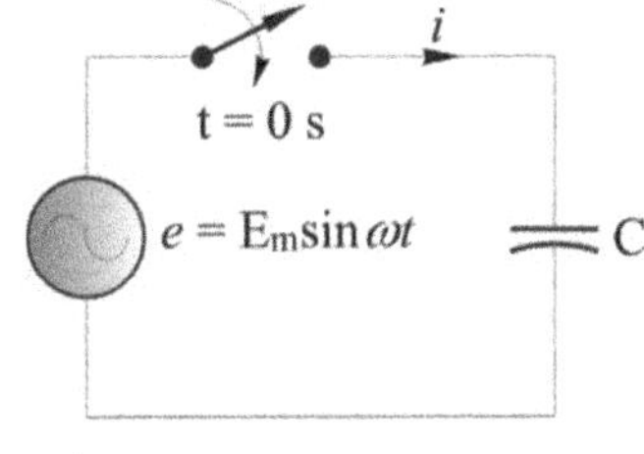

The effect of a capacitor is to delay a change in voltage and with a sinusoidal voltage, it causes the voltage to *lag* the current by 90° this is the same as saying that the current *leads* the voltage by 90°.

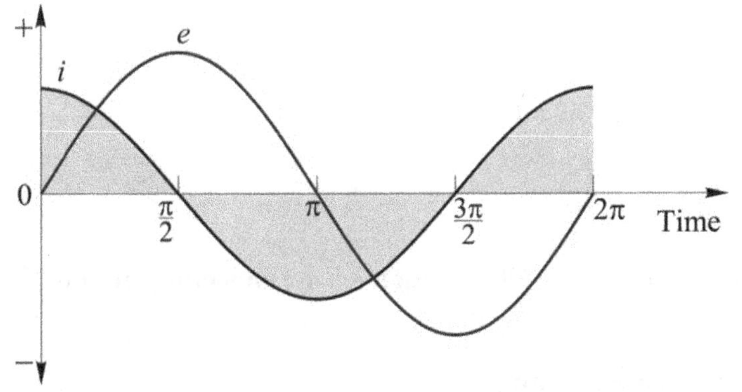

Phasor Diagram

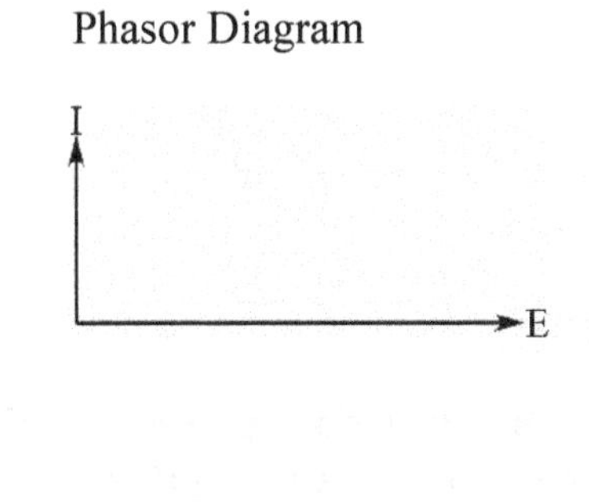

The magnitude of I is limited to a definite value. Therefore, the capacitor has an opposition to AC current flow. This is called the **capacitive reactance**, which has symbol $\mathbf{X_C}$ and is measured in **ohms**.

Ohm's Law can now be expressed in terms of capacitive reactance ⇨ $I = \dfrac{E}{X_C}$

X_C depends on capacitance and frequency ⇨ $X_C = \dfrac{1}{2\pi f C}$

As you can see from the expression above, reactive capacitance is inversely proportional to frequency, meaning that if the frequency increases, the capacitive reactance decreases, and if the frequency decreases, the capacitive reactance increases.

Example: How much current will be drawn from a 100-V, 60-Hz source when connected to a 100-μF capacitor?

$$X_C = \frac{1}{2\pi f C} = \frac{1}{377\ \text{rad/s} \times 100 \times 10^{-6}\ \text{F}} = 26.5\ \Omega \Rightarrow I = \frac{E}{X_C} = \frac{100\ \text{V}}{26.5\ \Omega} = \mathbf{3.77\ A}$$

EQUIPMENT

Power Supplies:
- Isolation transformer
- Function generator

Resistors: (5%, ½ W or higher)
- One each of 1-kΩ and 100-Ω

Instruments:
- DMM

Miscellaneous:
- Line cord

Other:
- Protoboard
- Hook-up Wire

Capacitors:
- One each of 1.0-μF, 0.47-μF, 0.22-μF, and 0.047-μF; all rated 100-V DC

PROCEDURE

Caution: High voltages are present in this experiment. Exhibit caution when connecting circuits and making measurements.

PART A: Effect of Capacitance on Capacitive Reactance

1. Before you start the lab, measure the output voltage of the transformer you will be using, as this will be the reference input voltage to the circuit.

2. Connect the circuit shown in Figure 17-1. Measure the current through the capacitor and record in Table 17-1.

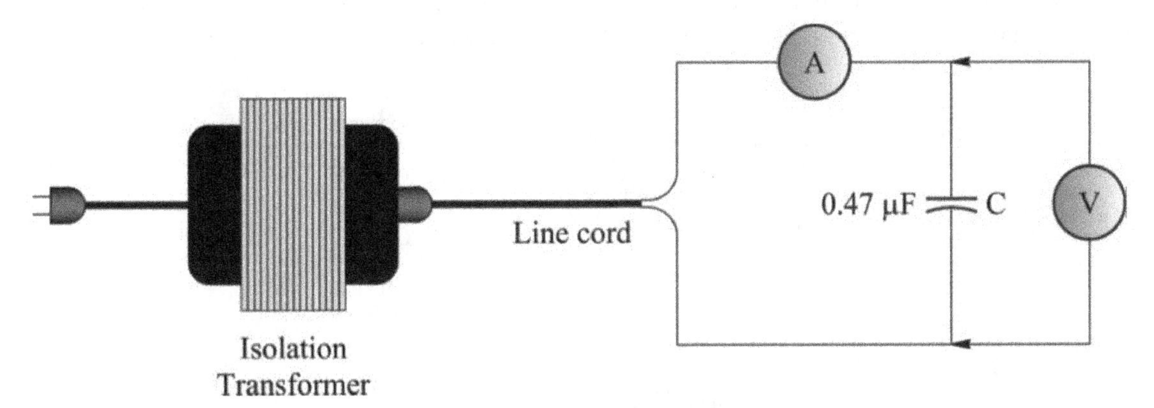

Figure 17-1

3. Repeat step 2 for capacitances of 1.0 μF and 0.22 μF.

   Calculate the capacitive reactance of each capacitor from Ohm's law, $X_C = V_C/I$ and also from the formula $X_C = 1/2\pi f C$. Record the values in Table 17-1.

PART B: Effect of Frequency on Capacitive Reactance

4. Connect the circuit in Figure 17-2.

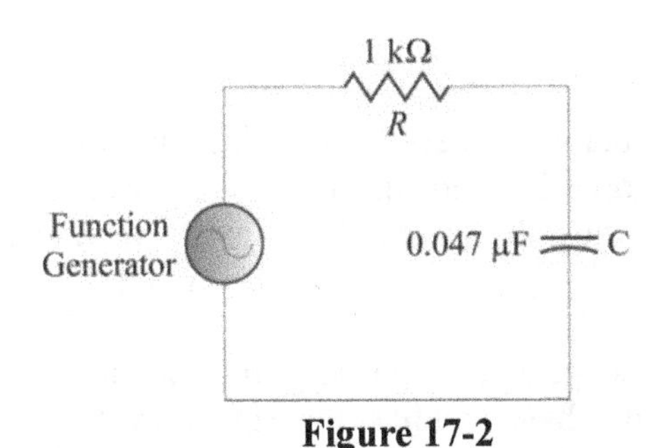

Figure 17-2

5. Turn **on** the function generator and set the frequency 1000 Hz. Using the DMM, adjust the output voltage of the function generator to10 V_{RMS}. The source voltage must be maintained at this value for the remainder of the experiment.

6. Using the DMM, measure the voltages across the capacitor and resistor and record the values in Table 17-2.

7. Adjust the frequency to the values shown in Table 17-2 and repeat step 7.

8. Calculate the current in the circuit from Ohm's Law $I = \frac{V_R}{R}$ for each frequency.

   Calculate, using $X_C = \frac{V_C}{I}$, the capacitive reactance for each frequency. Calculate the capacitive reactance from the formula $X_C = 1/2\pi f C$.

PART C: Phase Angle of Capacitor

9. Connect the circuit shown in Figure 17-3. The small resistance is used to derive a voltage proportional to the current; it has negligible effect on the phase angle.

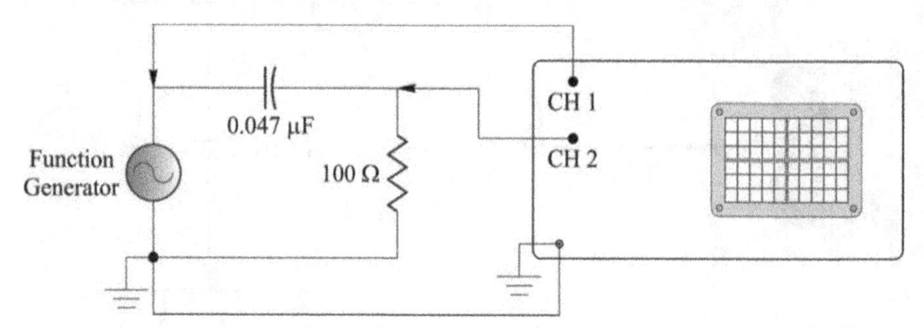

Figure 17-3

10. Switch **on** the oscilloscope. Set the trigger selector switch on the oscilloscope to CH 1. Set the mode switch to CH 1. Adjust the horizontal and vertical controls to obtain a centered trace.

11. Turn **on** the function generator. Set the frequency to 1 kHz and increase the output to maximum value.

12. The sine wave on the oscilloscope is the reference wave and is the applied voltage in the circuit. Using the volts/div controls, adjust the height of the sine wave to about six divisions peak-to-peak. Using the sec (time)/div controls and the variable sec (time)/div, adjust the sine wave so that one cycle occupies exactly ten horizontal divisions.

Each division will correspond to $\frac{360}{10} = 36°$

13. Set the mode switch to CH 2. Adjust the vertical control to obtain a centered trace. This sine wave is the voltage across the resistor and since resistor voltage is proportional to current, it also represents the current wave in this circuit. Using the volts/div controls adjust the height of the sine wave to about four divisions peak-to-peak.

14. Set the mode switch to Dual. Both waves will appear on the scope and the phase shift between voltage (CH 1) and current (CH 2) can easily be seen. You may have to adjust the trigger level control to reposition the CH 1 wave.

15. Measure the lead distance (d) between corresponding zero points on the voltage and current waves as shown in Figure 17-4.

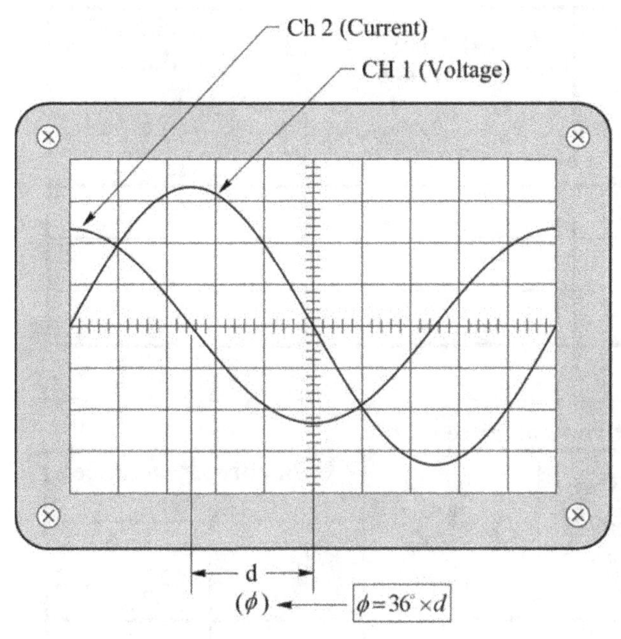

Figure 17-4

16. Determine the phase angle in degrees and record in Table 17-3.

RESULTS

Table 17-1 Reactance of a Capacitor

Capacitance C, (μF)	Capacitor Voltage V_C, (V)	Capacitor Current I_C, (mA)	Capacitive Reactance $X_C = V_C / I_C$ (Ω)	Capacitive Reactance $X_C = 1 / 2\pi f C$ (Ω)
0.47				
1.0				
0.22				

Table 17-2 Effect of Frequency on Capacitive Reactance

Frequency f, Hz	Source Voltage E (V)	Capacitor Voltage V_C (V)	Resistor Voltage V_R (V)	Current $I = V_R / R$ (A)	Capacitive Reactance X_C, Ω	
					V_C / I	$\frac{1}{2\pi f C}$
1000	10					
2000	10					
3000	10					
4000	10					
5000	10					
6000	10					
7000	10					
8000	10					

Table 17-3 Phase Angle of a Capacitor

Lead Distance d, (Divisions)	Phase Angle ϕ, (Degrees)

Capacitance Reactance vs. Frequency

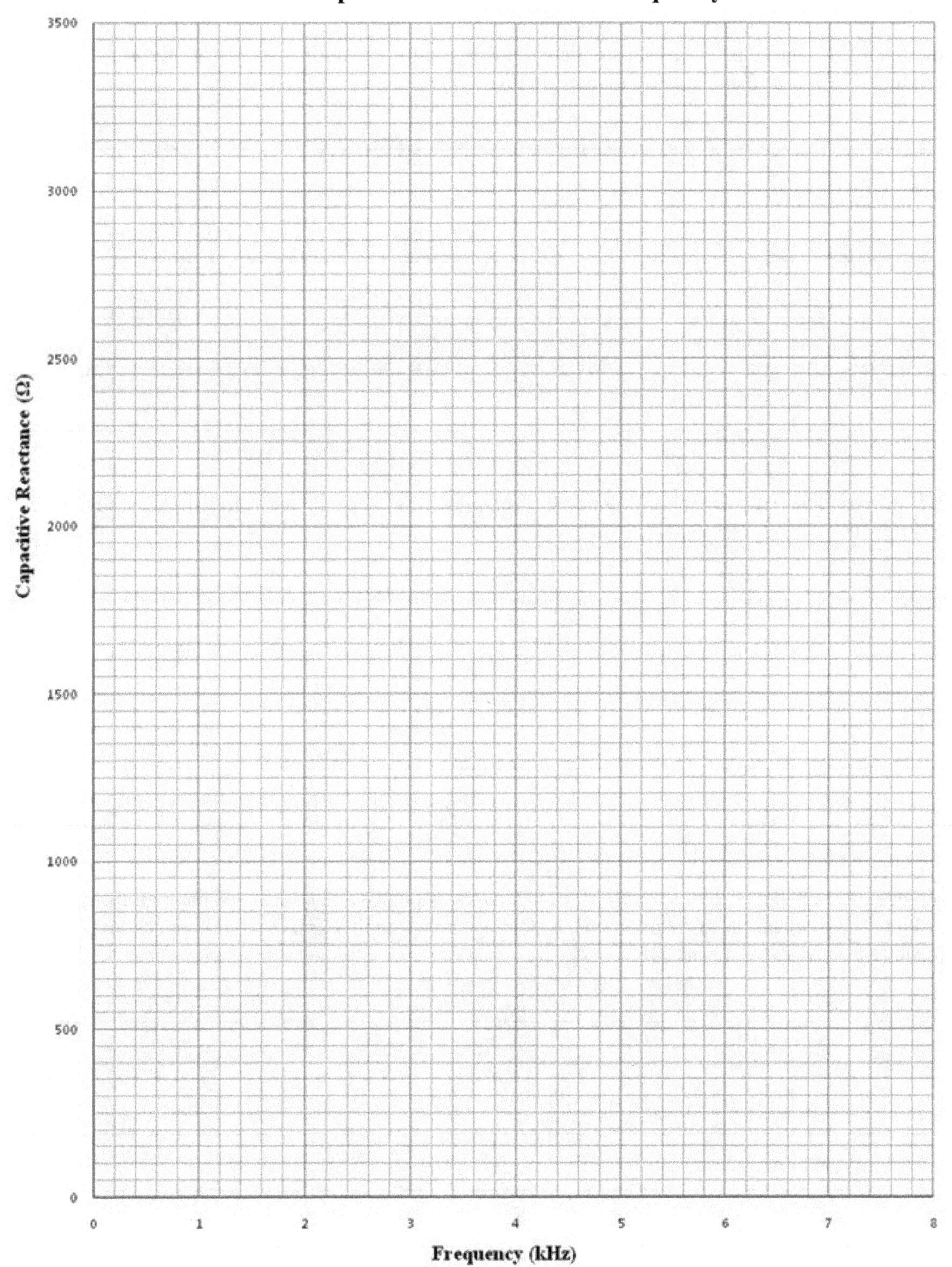

QUESTIONS

1. Do the results of your experiment substantiate the formula for capacitive reactance? Refer to your experimental results and explain any discrepancies.

2. How is the reactance of a capacitor affected by a change in **(a)** capacitance, and **(b)** frequency? Use your results to explain your answer.

3. On a piece of graph paper, plot a graph of frequency vs. capacitive reactance for the data from Table 17-2. Plot *frequency* on the horizontal axis. Referring to the graph, explain what happens to X_C if the frequency of the voltage is zero (DC).

4. Compare your measured value of phase angle for a pure capacitance with the theoretical value. What are the sources of error in this measurement?

____

____

____

____

____

5. A capacitor is connected to a 10-mV, 1.2-MHz source and a current of 50 μA flows. What is the value of capacitance?

____

____

____

____

____

EXPERIMENT

18

SERIES AND PARALLEL CAPACITORS

OBJECTIVES

1. To show experimentally that the total capacitance of series-connected capacitors is found from the reciprocal formula.

$$\frac{1}{C_T} = \frac{1}{C_1} + \frac{1}{C_2} + \frac{1}{C_3}$$

2. To show experimentally that the total capacitance of parallel-connected capacitors is the sum of the capacitances.

$$C_T = C_1 + C_2 + C_3$$

DISCUSSION

Capacitors in Parallel

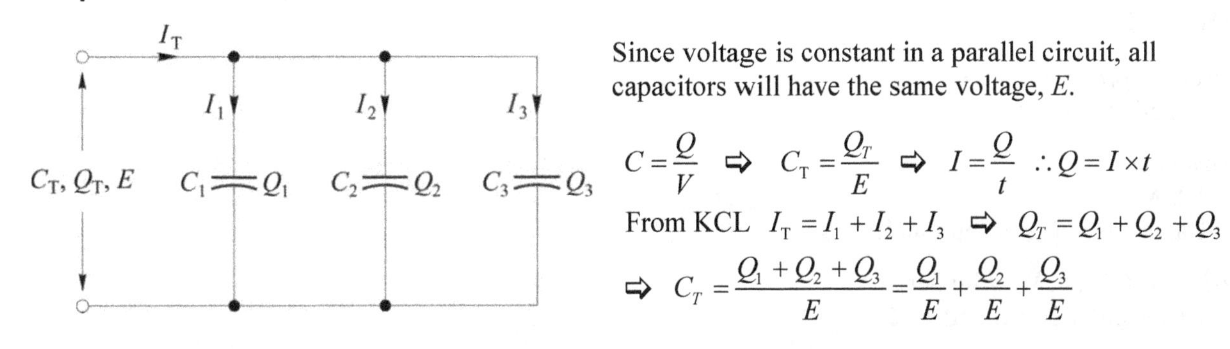

Since voltage is constant in a parallel circuit, all capacitors will have the same voltage, E.

$$C = \frac{Q}{V} \Rightarrow C_T = \frac{Q_T}{E} \Rightarrow I = \frac{Q}{t} \quad \therefore Q = I \times t$$

From KCL $I_T = I_1 + I_2 + I_3 \Rightarrow Q_T = Q_1 + Q_2 + Q_3$

$$\Rightarrow C_T = \frac{Q_1 + Q_2 + Q_3}{E} = \frac{Q_1}{E} + \frac{Q_2}{E} + \frac{Q_3}{E}$$

$C_T = C_1 + C_2 + C_3$ ⇨ Capacitors in parallel are added directly.

Capacitors in Series

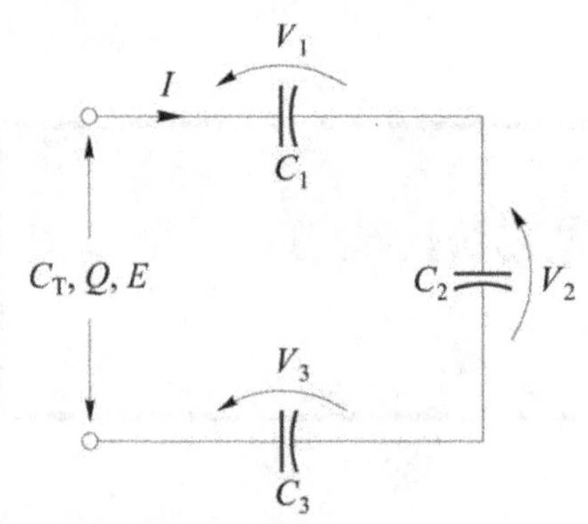

Since current is constant in a series circuit, it follows that the charge on each capacitor will be the same, Q coulombs.

$$C_T = \frac{Q}{E} \Rightarrow E = V_1 + V_2 + V_3$$

$$\therefore C_T = \frac{Q}{V_1 + V_2 + V_3} \Rightarrow \frac{1}{C_T} = \frac{V_1 + V_2 + V_3}{Q} = \frac{V_1}{Q} + \frac{V_2}{Q} + \frac{V_3}{Q}$$

$$\frac{1}{C_T} = \frac{1}{C_1} + \frac{1}{C_2} + \frac{1}{C_3}$$ ⇨ Series capacitors combine using the reciprocal formula.

Example: A 50-µF capacitor and a 100-µF capacitor are connected in series to a 150-V source. Calculate the total capacitance and the voltage drop across each capacitor.

$$C_T = \frac{50\,\mu\text{F} \times 100\,\mu\text{F}}{150\,\mu\text{F}} = \mathbf{33.3\,\mu F}$$

$$C = \frac{Q}{V} \Rightarrow Q = CV = 33.3 \times 10^{-6}\,F \times 150\,\text{V} = 5.0\,\text{mC}$$

$$V = \frac{Q}{C} \Rightarrow V_1 = \frac{Q}{C_1} = \frac{5.0 \times 10^{-3}\,\text{C}}{50 \times 10^{-6}\,\text{F}} = \mathbf{100\,V} \qquad V_2 = \frac{Q}{C_2} = \frac{5.0 \times 10^{-3}\,\text{C}}{100 \times 10^{-6}\,\text{F}} = \mathbf{50\,V}$$

EQUIPMENT

Power Supplies:
- Function generator

Instruments:
- DMM
- Capacitance Meter (optional)

Resistors: (5%, ½ W or higher)
- One each of 100-Ω and 1-kΩ

Capacitors:
- One each of 0.22-µF, 0.47-µF, and 1.0-µF; all rated 100-V DC

Other:
- Protoboard
- Hook-up wire

PROCEDURE

PART A: Capacitors in Series

Note: If a capacitance meter is not available or your DMM does not measure capacitance, omit steps 1 and 5.

1. Connect the circuit in Figure 18-1 and measure the total capacitance of the series combination with a capacitance meter or your DMM and record the value in Table 18-1.

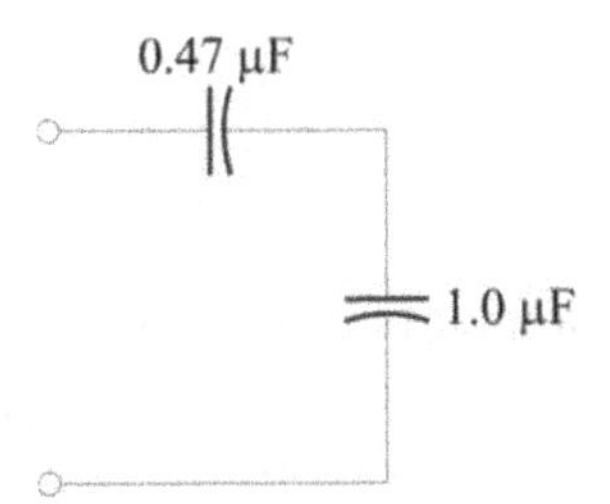

Figure 18-1

2. Connect the circuit shown in Figure 18-2. Use the DMM on the AC current function for the ammeter.

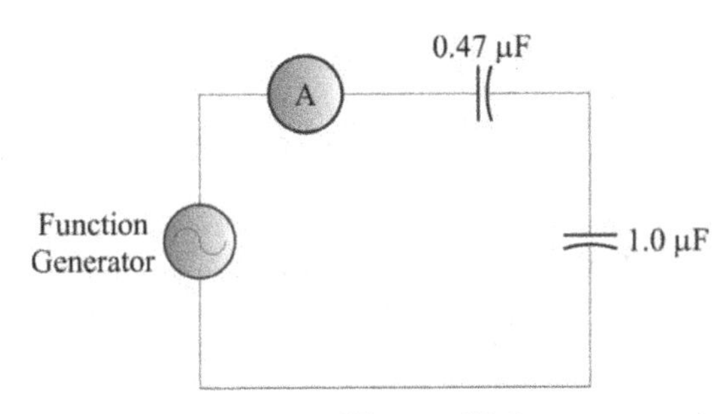

Figure 18-2

3. Switch **on** the function generator and set the frequency to 1000 Hz. Increase the source voltage until the voltage across the series circuit is 5 V_{RMS}. Measure the total current and record the value in Table 18-1.

4. Calculate the total capacitive reactance (X_{CT}), from E/I. Calculate total capacitance of the series combination from the reciprocal formula and record all values in Table 18-1.

5. Add a 0.22 μF capacitor to the series combination and repeat steps 1 to 4.

PART B: Capacitors in Parallel

6. Connect the circuit in Figure 18-3 and measure the total capacitance of the parallel combination with a capacitance meter or the capacitance function on you DMM and record the value in Table 18-2.

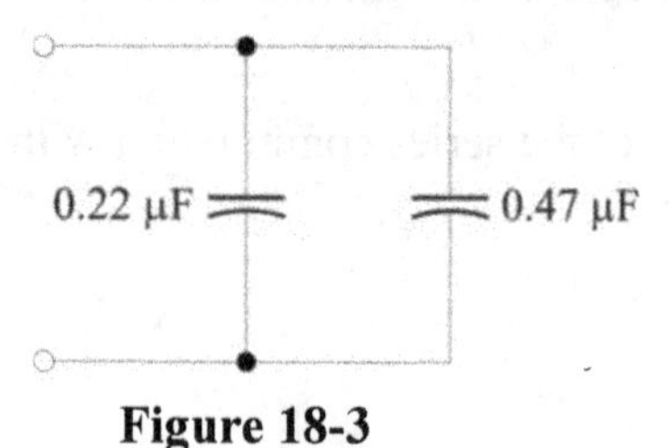

Figure 18-3

7. Connect the circuit in Figure 18-4. Use the DMM on the AC current function for the ammeter.

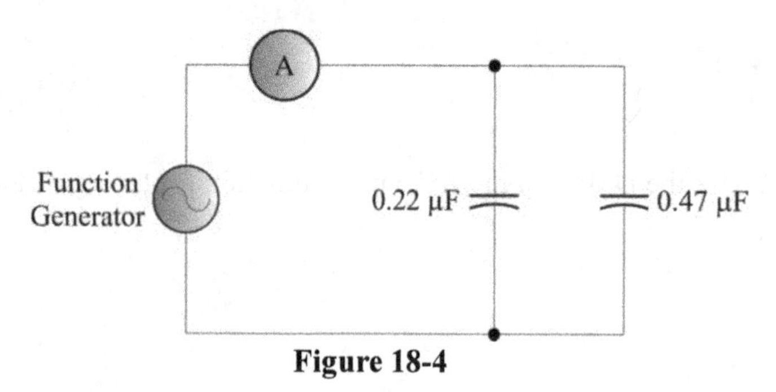

Figure 18-4

8. Switch **on** the function generator and set the frequency to 1000 Hz. Increase the source voltage until the voltage across the series circuit is 5 V_{RMS}. Measure the total current and record the value in Table 18-2.

9. Calculate the total capacitive reactance, X_{CT}, from E/I. Calculate total capacitance of the parallel combination from the formula and record all values in Table 18-2.

10. Connect a 1.0 μF capacitor to the parallel combination and repeat steps 8 to 9.

RESULTS

Table 18-1 Capacitors in Series

Capacitors C_T (μF)	C_T (meas.) (μF)	Circuit Voltage V_{AD} (V)	V_R (V)	I (mA)	X_{CT} (Ω)	C_T (calc.) (Ω)
1.0, 0.47		5				
1.0, 0.47, 0.22		5				

Table 18-2 Capacitors in Parallel

Capacitors C_T (μF)	C_T (meas.) (μF)	Circuit Voltage V_{AD} (V)	V_R (V)	I (mA)	X_{CT} (Ω)	C_T (calc.) (Ω)
1.0, 0.47		5				
1.0, 0.47, 0.22		5				

QUESTIONS

1. Compare the values of C_T from Table 18-1. What are possible sources of error?

2. Referring to your results, discuss the effect of adding series capacitors to total capacitance and total current. Explain the reason for this effect.

3. By referring to your results, discuss the effect of adding parallel capacitors to total capacitance and total current. Explain the reason for this effect.

4. What is the total capacitance of a 0.005-μF, 10-nF, and 2000-pF capacitor in series?

____

____

____

____

____

____

5. A 0.05-μF capacitor is connected in parallel with a 10-nF capacitor. How much capacitance must be connected in parallel with this combination if the total capacitance is to be 100-nF?

____

____

____

____

____

____

d) [illegible]

e) [illegible]

EXPERIMENT

19

SERIES AND PARALLEL INDUCTORS

OBJECTIVES

1. To investigate the effect that DC current has on the inductance of a choke.
2. To show that the total inductance of series-connected inductors is the sum of the inductances.

$$L_T = L_1 + L_2 + L_3$$

3. To show that the total inductance of parallel-connected inductors is found from the reciprocal formula.

$$\frac{1}{L_T} = \frac{1}{L_1} + \frac{1}{L_2} + \frac{1}{L_3}$$

DISCUSSION

Inductances in Series

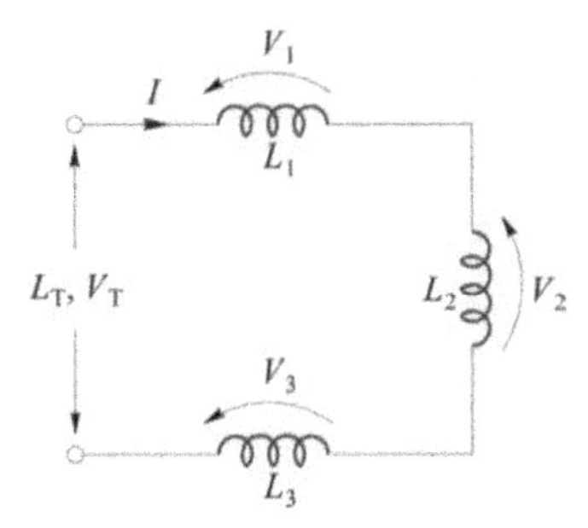

For an inductor $V = L\left(\frac{dI}{dt}\right)$ ⇨ $L = \frac{V}{\left(\frac{dI}{dt}\right)}$

In series connected inductors, the same current flows through each one and therefore they are all subjected to the same rate of change of current $\frac{dI}{dt}$.

$$L_T = \frac{V_T}{\left(\frac{dI}{dt}\right)} = \frac{V_1 + V_2 + V_3}{\left(\frac{dI}{dt}\right)} \quad ⇨ \quad L_T = \frac{V_1}{\left(\frac{dI}{dt}\right)} + \frac{V_2}{\left(\frac{dI}{dt}\right)} + \frac{V_3}{\left(\frac{dI}{dt}\right)}$$

$$L_T = L_1 + L_2 + L_3$$

This relationship can be verified experimentally by applying a voltage to the series connected inductors and measuring the current flowing.

Inductances in Parallel

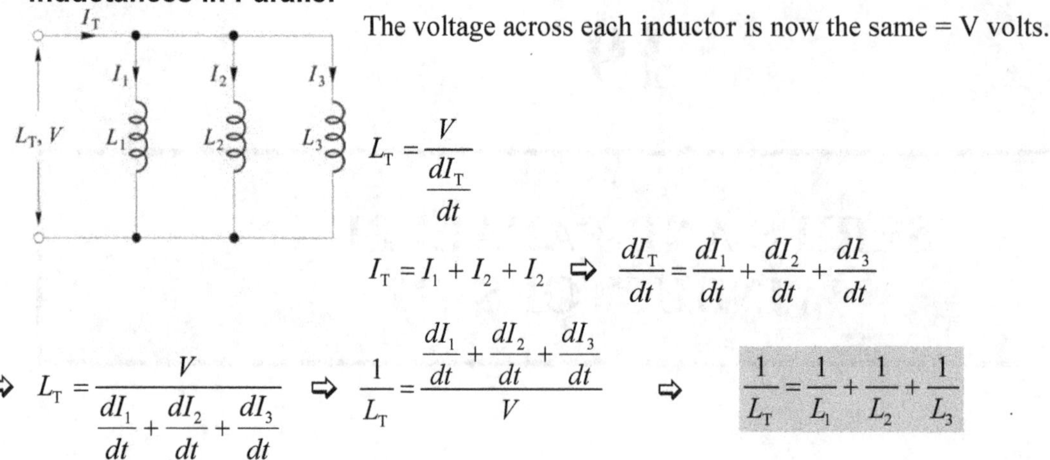

The voltage across each inductor is now the same = V volts.

$$L_T = \frac{V}{\frac{dI_T}{dt}}$$

$$I_T = I_1 + I_2 + I_2 \Rightarrow \frac{dI_T}{dt} = \frac{dI_1}{dt} + \frac{dI_2}{dt} + \frac{dI_3}{dt}$$

$$\Rightarrow L_T = \frac{V}{\frac{dI_1}{dt} + \frac{dI_2}{dt} + \frac{dI_3}{dt}} \Rightarrow \frac{1}{L_T} = \frac{\frac{dI_1}{dt} + \frac{dI_2}{dt} + \frac{dI_3}{dt}}{V} \Rightarrow \frac{1}{L_T} = \frac{1}{L_1} + \frac{1}{L_2} + \frac{1}{L_3}$$

Parallel inductors combine in the same manner as parallel resistors. This can be verified experimentally in the same manner as for the series connection.

EQUIPMENT

Power Supplies:
- Function generator

Instruments:
- DMM

Resistors: (5%, ½ W or higher)
- One each of 100-Ω and 1-kΩ

Inductors:
- RF inductors 25 mH, 50-mH, and 100-mH

Other:
- Protoboard
- Hook-up wire

PROCEDURE

PART A: Inductors in Series

1. Connect the circuit in Figure 19-1. Use the DMM on AC current function for the ac ammeter.

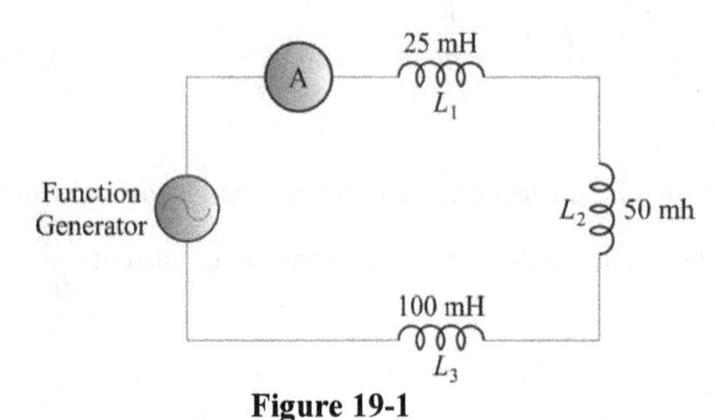

Figure 19-1

2. Switch **on** the function generator and set the frequency to 2000 Hz. Increase the source voltage until the voltage across the series circuit is 5V $_{RMS}$. Measure the total current and record the value in Table 19-1.

3. Calculate the total inductive reactance, X_{LT}, from. $\frac{E}{I}$. Calculate total inductance of the series combination from $L_T = \frac{X_{LT}}{2\pi f}$ and record all values in Table 19-1.

PART B: Inductors in Parallel

4. Connect the circuit in Figure 19-2.

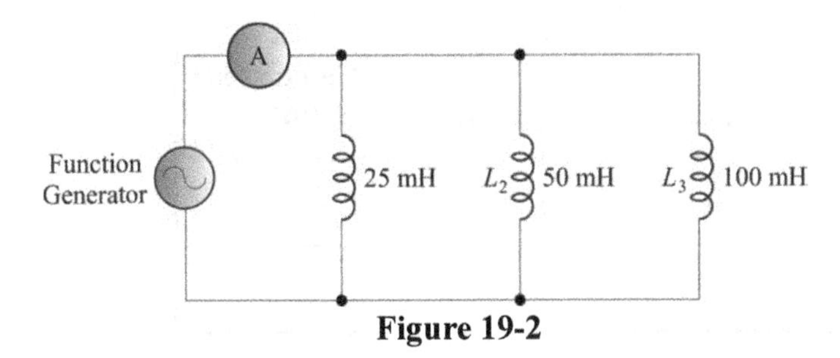

Figure 19-2

5. Switch **on** the function generator and set the frequency to 2000 Hz. Increase the source voltage until the voltage across the series circuit is 5 V $_{RMS}$. Measure the total current and record the value in Table 19-1.

6. Calculate the total inductive reactance (X_{LT}), from E/I. Calculate total inductance of the parallel combination from $L_T = \dfrac{X_{LT}}{2\pi f}$ and record all values in Table 19-1.

PART C: Inductor Circuit Design

7. Design and sketch a circuit using the inductors from this experiment which have a total inductance of 60 mH. A series-parallel combination will be necessary. Show your calculations.

8. Connect your designed circuit to the function generator and set the frequency to 2000 Hz. Increase the source voltage until the voltage across the series circuit is 5 V. Measure the total current and record the value in Table 19-2.

9. Calculate the total inductive reactance (X_{LT}), from E/I. Calculate total inductance of the circuit from $L_T = \dfrac{X_{LT}}{2\pi f}$ and record in Table 19-2.

RESULTS

Table 19-1 Inductors in Series and Parallel

Circuit	Circuit Voltage, V_{AD} (V)	I_T (mA)	X_T (Ω)	L_T (mH)	L_T (Calculated) (mH)
Series	5				
Parallel	5				

PART C: Design Circuit

Table 19-2 Inductor Circuit Design

L_T Design (mH)	Circuit Voltage, V_{AD} (V)	I_T (mA)	X_T (Ω)	L_T (mH)
60	5			

QUESTIONS

1. Compare the measured values of L_T to the values calculated from the combination formulas for the series and parallel circuits. What are possible sources of error?

2. For part (C), calculate the % error between your designed and measured total inductance. What are the major sources of error?

3. What is the total inductance of a 500-μH, 120-mH, and 0.05-H inductor in series?

4. A 50-mH inductor is connected in parallel with a 30-mH inductor. How much inductance must be connected in parallel with this combination if the total inductance is to be 0.01-H?

EXPERIMENT

20

INDUCTIVE REACTANCE

OBJECTIVES

1. To observe the behavior of inductance in AC and DC circuits.
2. To prove that $X_L = 2\pi f L$.

DISCUSSION

Inductive Reactance

When AC voltage is applied across a pure inductance, the sinusoidal variations of the voltage cause the inductance to *oppose* the flow of AC current and limit it to a definite value. This opposition that an inductor has to AC current flow is called **inductive reactance**, which has symbol X_L and is measured in **ohms**.

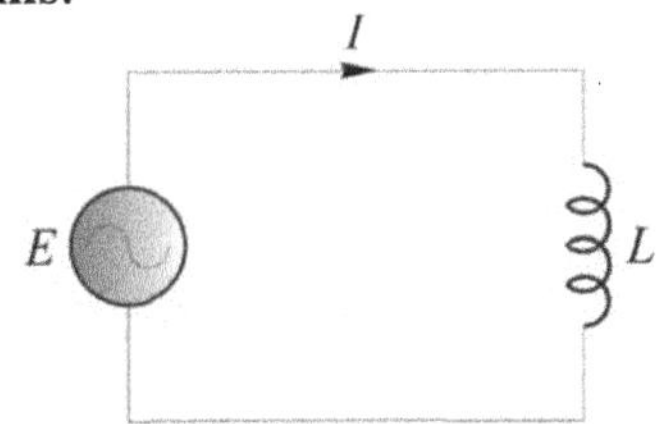

Ohm's Law can now be rewritten for this circuit as $I = \frac{E}{X_L}$

The inductive reactance depends on two factors:

1. The **inductance (L)** of the coil in henrys;
2. The **frequency (f)** of the supply voltage.

The formula for inductive reactance is $X_L = 2\pi f L$.

Practical Inductor

Since an inductor is a coil of wire, it will always have a certain amount of resistance (r_{coil}) due to the resistance of the wire as shown below. Therefore it is not possible to have a pure inductance because of this inherent resistance. A practical inductor will always have a small amount of resistance along with its inductive reactance.

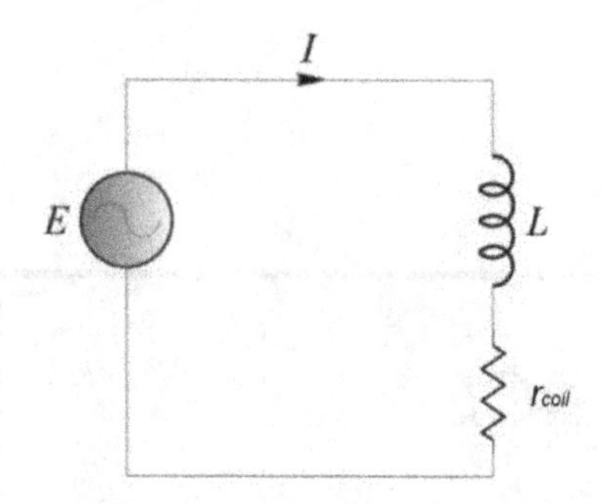

When an inductor is connected to a DC voltage source, only the resistance (r_{coil}) opposes the current and large currents can flow. When an AC source is connected, the inductive reactance, which is much higher than the resistance, will limit the current to smaller values.

Example: How much current will flow through a 0.2-H inductor if it is connected to a 250-V, 60-Hz source?

$$X_L = 2\pi f L = 2\pi \times 60\,\text{Hz} \times 0.2\,\text{H} = 75.4\,\Omega$$

$$I = \frac{E}{X_L} = \frac{250\,\text{V}}{75.4\,\text{A}} = \mathbf{3.32A}$$

EQUIPMENT

Power Supplies:
- Function generator
- 0-24 V regulated DC supply

Instruments:
- DMM
- Oscilloscope
- Protoboard

Resistors: (5%, ½ W or higher)
- One each of 1-kΩ and 100-Ω

Inductors:
- One each 25mH , 50 mH, and 100 mH inductor

Other:
- Hook-up Wire

PROCEDURE

PART A: DC and AC Currents in Inductance

1. Measure the *resistance* of the 100-mH inductor and record this value in Table 20-1.

2. Connect the circuit shown in Figure 20-1 using the DC power supply.

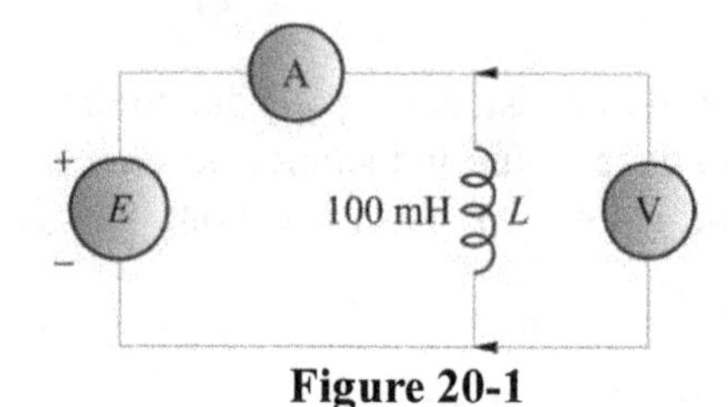

Figure 20-1

3. Turn **on** the power supply and adjust the source voltage until the voltage across the inductor is 5 V. Measure the current flowing in the inductor and record this value in Table 20-1.

4. Connect the 100-mH inductor to the Function Generator as shown in Figure 20-2.

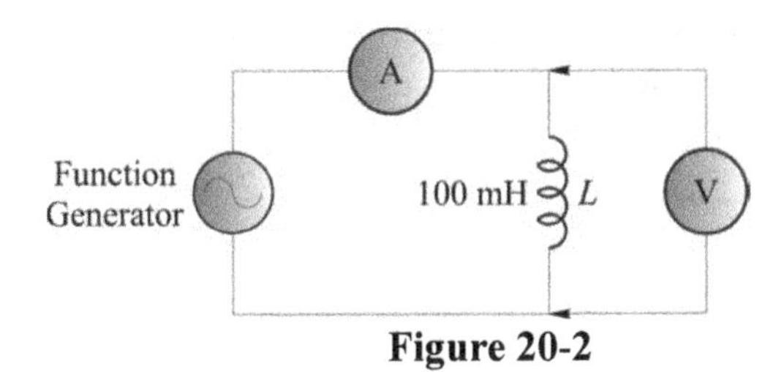

Figure 20-2

5. Turn **on** the Function Generator and adjust frequency to 3000 Hz. Adjust the source voltage until the voltage across the inductor is 5 V_{RMS}. Measure and record the AC current flowing in the inductor.

   Calculate the inductive reactance from $X_L = \frac{V_L}{I}$ and record in Table 20-2.

6. Repeat step 5 for inductances of 25 and 50 mH.

PART B: Effect of Frequency on Inductive Reactance

7. Connect the circuit shown in Figure 20-3.

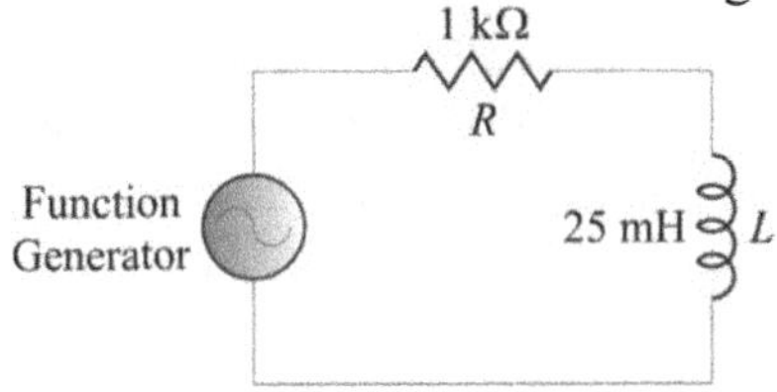

Figure 20-3

8. Turn **on** the function generator and set the frequency 1000 Hz. Using the DMM, adjust the output voltage of the function generator to 7 V_{RMS}. The source voltage must be maintained at this value for the remainder of the experiment.

9. Using the DMM measure the voltages across the inductor and resistor and record the values in Table 20-3.

10. Adjust the frequency to the values shown in Table 20-3 and repeat step 8.

11. Calculate the current in the circuit from Ohm's Law $I = \frac{V_R}{R}$ for each frequency.

12. Calculate, using $X_L = \frac{V_L}{I}$, the inductive reactance for each frequency.

PART C: Phase Angle of an Inductor

13. Connect the circuit shown in Figure 20-4. The small resistance is used to derive a voltage proportional to the current; it has negligible effect on the phase angle.

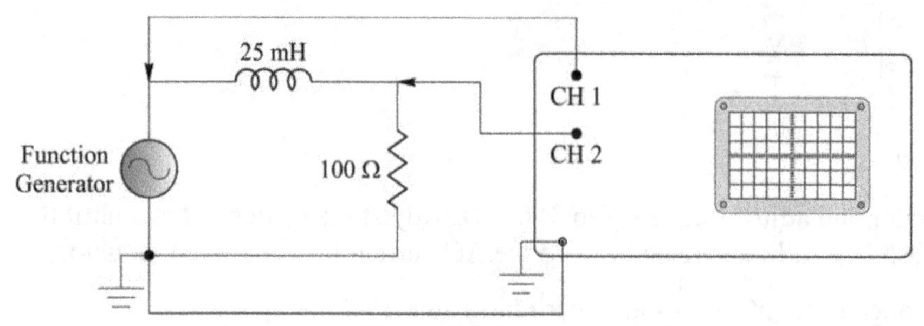

Figure 20-4

14. Switch **on** the oscilloscope. Set the trigger selector switch on the oscilloscope to CH 1. Set the mode switch to CH 1. Adjust the horizontal and vertical controls to obtain a centred trace.

15. Plug the function generator into the isolation transformer and switch it on. Set the frequency to 15 kHz and increase the output to maximum value.

16. The sine wave on the oscilloscope is the reference wave and is the applied voltage in the circuit. Using the volts/div controls, adjust the height of the sine wave to about six divisions peak-to-peak. Using the sec (time)/div controls and the variable sec (time)/div, adjust the sine wave so that one cycle occupies exactly ten horizontal divisions.

Each division will correspond to $\frac{360}{10} = 36°$

Set the mode switch to CH 2. Adjust the vertical control to obtain a centered trace. This sine wave is the voltage across the resistor and since resistor voltage is proportional to current, it also represents the current wave in this circuit. Using the volts/div controls, adjust the height of the sine wave to about four divisions peak-to-peak

17. Set the mode switch to dual. Both waves will appear on the scope and the phase shift between voltage (CH 1) and current (CH 2) can easily be seen. You may have to adjust the trigger level control to reposition the CH 1 wave.

18. Measure the lag distance (d) between corresponding zero points on the voltage and current waves as shown in Figure 20-5.

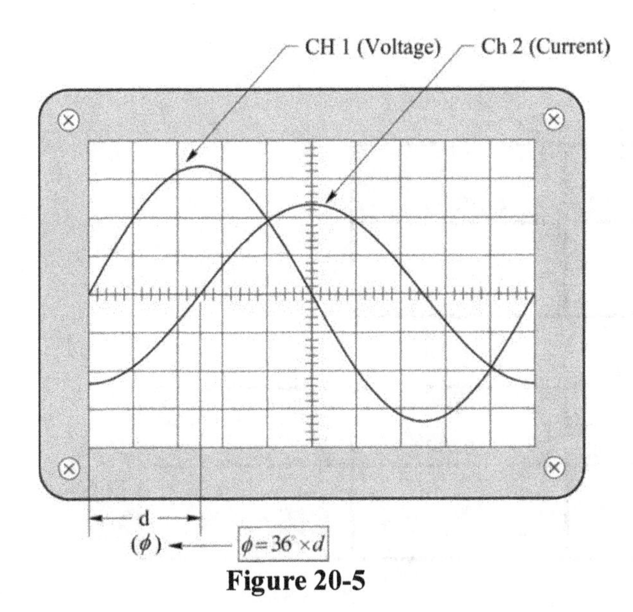

Figure 20-5

19. Determine the phase angle in degrees and record in Table 20-4.

RESULTS

Table 20-1 DC and AC Currents in Inductance

L (mH)	r_{coil} (Ω)	V_L (V)	I (mA)
100		5	

Table 20-2 Effect of Frequency on Reactance

L (mH)	r_{coil} (Ω)	V_L (V)	I (mA)	X_L (Ω)
100		5		

Table 20-3 Effect of Frequency on Reactance

Frequency f, Hz	E (V)	V_L (V)	V_R (V)	$I = V_R/R$ (A)	X_L (Ω)	
					V_L/I	$2\pi f L$
1000	7					
2000	7					
3000	7					
4000	7					
5000	7					

Table 20-4 Phase of an Inductor

Lag Distance d, (Div)	Phase angle ϕ, (Deg)

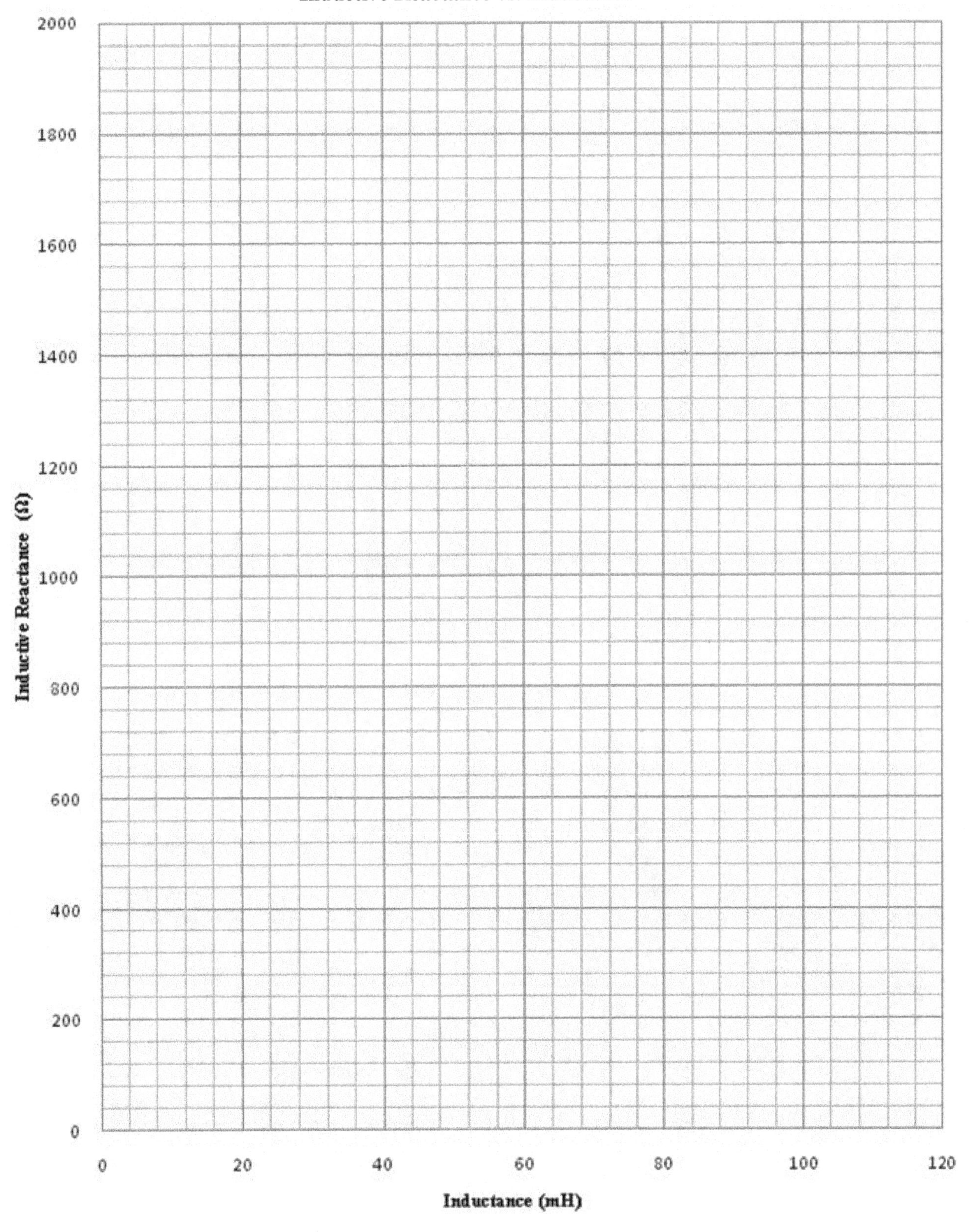
Inductive Reactance vs. Inductance
2000
1800
1600
1400
1200
1000
800
600
400
200
0
Inductive Reactance (Ω)
0
20
40
60
80
100
120
Inductance (mH)

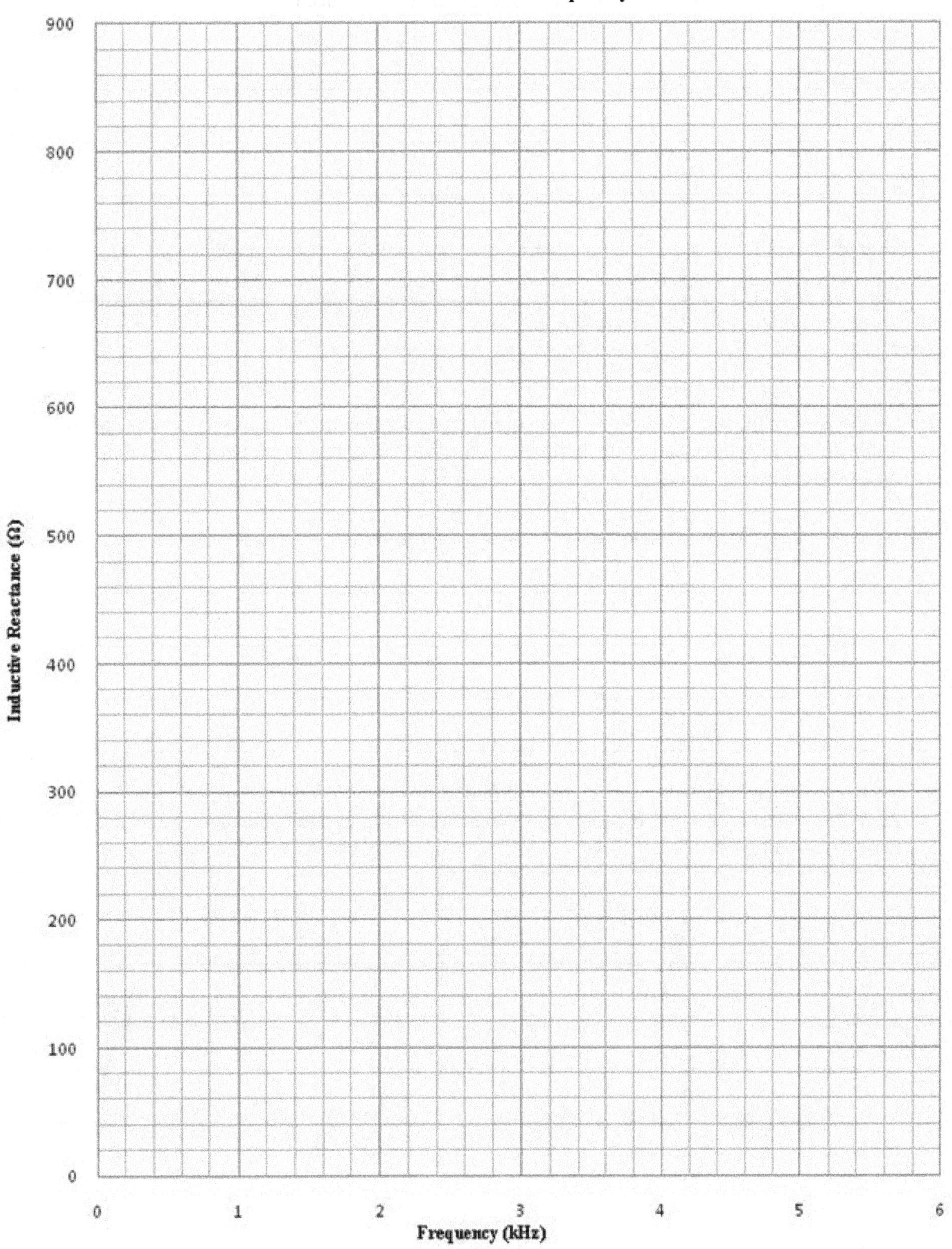
Inductive Reactance vs. Frequency
900
800
700
600
500
400
300
200
100
0
Inductive Reactance (Ω)
0
1
2
3
4
5
6
Frequency (kHz)

QUESTIONS

1. Referring to data in Table 20-1, describe the effect that inductance has on AC current compared with DC current. Explain why this happens.

2. On the graph paper included, plot a graph of inductance vs. inductive reactance for the data from Table 20-1. Plot *inductance* on the horizontal axis. What does this graph tell you about the relationship between X_L and L?

3. On the graph paper included, plot a graph of frequency vs. inductive reactance for the data from Table 20-3. Plot *frequency* on the horizontal axis. What does this graph tell you about the relationship between X_L and f?

4. How does the measured value of phase angle compare to the theoretical value for an inductor? Comment on possible sources of error.

5. At what frequency will a 200-μH inductance have a reactance of 20 Ω? How much current will flow if this inductance is now connected to a 30-V, 10-kHz source?

4. [illegible]

5. [illegible]

EXPERIMENT

21

THE SERIES *RL* CIRCUIT

OBJECTIVES

1. To verify the expression for total impedance of a series *RL* circuit.

$$Z = \sqrt{R^2 + X_L^2}$$

2. To determine the phase angle between voltage and current in a series *RL* circuit.
3. To verify the voltage relationship of a series *RL* circuit.

$$E = \sqrt{V_R^2 + V_L^2}$$

DISCUSSION

Series *RL* Circuit

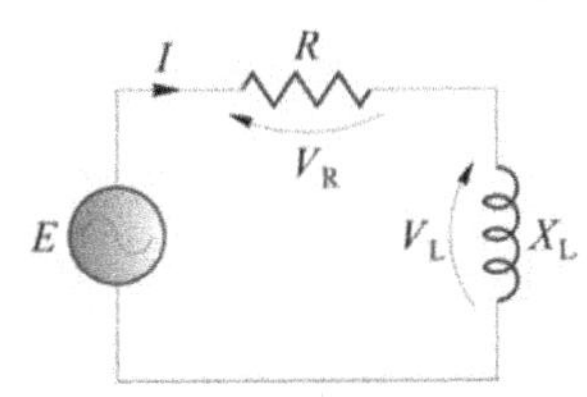

Kirchhoff's voltage law must hold true for this circuit and can be written as

⇨ $\vec{E} = \vec{V}_R + \vec{V}_L$.

The voltages are now **vector** quantities and using current as the reference phasor, we obtain the following phasor diagram:

$$|E| = \sqrt{V_L^2 + V_R^2}$$

The angle between voltage and current is called the **phase angle (ϕ)** of the circuit.

$$\phi = \tan^{-1}\left(\frac{V_L}{V_R}\right)$$

The resistance and inductive reactance combine to form a total opposition to AC current flow. This total opposition is defined as **impedance**, symbolized by Z and measured in **ohms**.

The basic Ohm's law equation for any AC circuit can now be written as $I = \frac{E}{Z}$.

$$I = \frac{E}{Z} \Rightarrow Z = \frac{E}{I} \quad \text{but} \quad E = \sqrt{V_R^2 + V_L^2}$$

$$Z = \frac{\sqrt{V_R^2 + V_L^2}}{I} = \sqrt{\frac{V_R^2 + V_L^2}{I^2}} \Rightarrow Z = \sqrt{\frac{V_R^2}{I^2} + \frac{V_L^2}{I^2}} = \sqrt{\left(\frac{V_R}{I}\right)^2 + \left(\frac{V_L}{I}\right)^2}$$

$$\text{but} \quad \frac{V_R}{I} = R \quad \text{and} \quad \frac{V_L}{I} = X_L \Rightarrow Z = \sqrt{R^2 + X_L^2}$$

This relationship takes the form of a Pythagorean equation and therefore the relationship can be represented by a right-angle triangle called the **impedance triangle**.

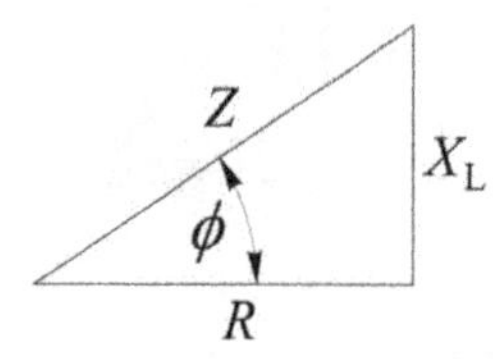

From the triangle, it can be seen that the phase angle of the circuit can be written as:

$$\phi = \tan^{-1}\left(\frac{X_L}{R}\right)$$

Example: A circuit has a resistance of 50-Ω and inductance of 0.2-H. It is connected to a 220-V, 50-Hz source. Calculate impedance, current, and phase angle.

$$X_L = 2\pi f L = 2\pi \times 50\,\text{Hz} \times 0.2\,\text{H} = 62.8\,\Omega$$

$$\Rightarrow Z = \sqrt{R^2 + X_L^2} = \sqrt{50^2 + 62.8^2} = \mathbf{80.3\,\Omega}$$

$$\Rightarrow I = \frac{E}{Z} = \frac{220\,\text{V}}{80.3\,\Omega} = \mathbf{2.74A}$$

$$\Rightarrow \phi = \tan^{-1}\frac{X_L}{R} = \tan^{-1}\frac{62.8}{50} = \tan^{-1}1.26 = \mathbf{51.5°}$$

EQUIPMENT

Power Supplies:
- Function generator

Instruments:
- DMM
- Oscilloscope

Resistors: (5%, ½ W or higher)
- One each of 1500-Ω, 1000-Ω, 470-Ω, and 560-Ω

Inductors:
- One 25 and 50 mH inductor

Other:
- Protoboard
- Hook-up Wire

PROCEDURE

PART A: Impedance of a Series *RL* Circuit

1. Connect the circuit shown in Figure 21-1. Set the function generator to a frequency to 2 kHz and increase the output to 5 V RMS (measured by DMM).

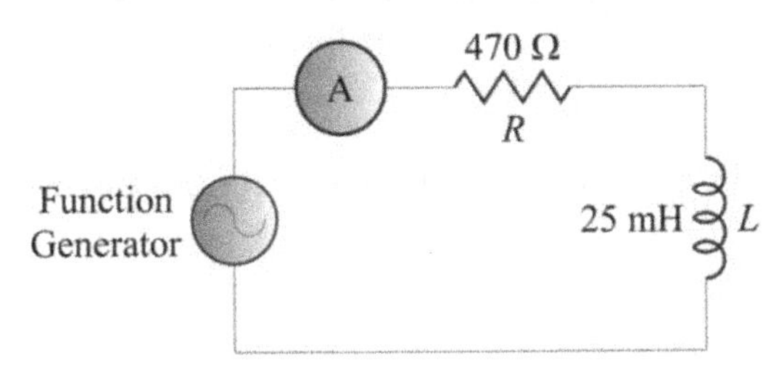

Figure 21-1

2. Measure current flowing in the circuit and record in Table 21-1. Measure the voltage across the inductor and record in Table 21-1.

3. Calculate X_L from the measured values of V_L and I. Calculate total impedance from E/I. Record the values in Table 21-1.

4. Replace the resistor with the 1.0-kΩ resistor and the inductor with a 50-mH inductor and repeat steps 2 and 3.

PART B: Phase and Voltage Relationships of a Series *RL* Circuit

5. Connect the circuit shown in Figure 21-2.

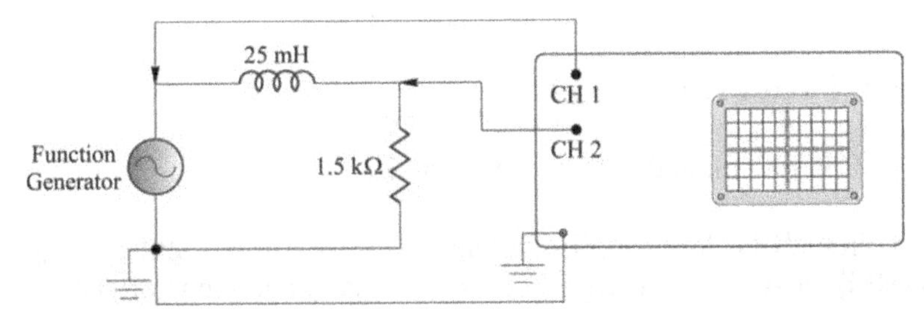

Figure 21-2

6. Switch **on** the oscilloscope. Set the trigger selector switch on the oscilloscope to CH 1. Adjust the horizontal and vertical controls to obtain a centered trace.

7. Set the frequency of the function generatorto 10 kHz and increase the output to 5 V RMS (measured by DMM).

8. Set the mode switch to CH 1. The sine wave on the oscilloscope is the reference wave and is the applied voltage in the circuit. Using the volts/div controls, adjust the height of the sine wave to about six divisions peak-to-peak. Using the sec (time)/div controls and the variable sec (time)/div, adjust the sine wave so that one cycle occupies exactly ten divisions. Each division will correspond to $360/10 = 36°$.

9. Set the mode switch to CH 2. Adjust the vertical control to obtain a centered trace. This sine wave is the voltage across the resistor and since resistor voltage is proportional to current, it also represents the current wave in this circuit. Using the volts/div controls, adjust the height of the sine wave to about four divisions peak-to-peak.

10. Set the mode switch to Dual. Both waves will appear on the scope and the phase shift between voltage (CH 1) and current (CH 2) can easily be seen. You may have to adjust the trigger level control to reposition the CH 1 wave.

11. Measure the lag distance (d) between corresponding zero points on the voltage and current waves as shown in Figure 21-3.

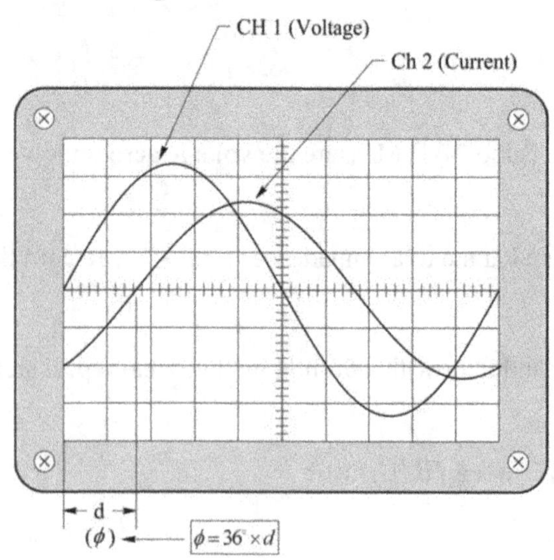

Figure 21-3

12. Determine the phase angle in degrees and record in Table 21-2.

13. Replace the 1.5-kΩ resistor with the 560-Ω resistor and repeat steps 8 to 12.

14. Disconnect the oscilloscope from the circuit. Check the applied voltage to verify that it is still 5 V_{RMS} and adjust if necessary. Using the DMM, measure the voltages across the resistor and inductor and record in Table 21-3.

15. Calculate the current in the circuit using Ohm's law from the measured values of V_R and R. Calculate the inductive reactance, X_L, from the equation $X_L = 2\pi f L$. Record these in Table 21-3.

16. Calculate the phase angle of the circuit from $\phi = \tan^{-1}\left(\frac{X_L}{R}\right)$. Record this value in Table 21-3.

17. Using the vector relationship, calculate the applied voltage from the measured values of V_L and V_R.

$$E = \sqrt{V_R^2 + V_L^2}$$

18. Repeat steps 14 to 17 with the 1.5-kΩ resistor.

RESULTS

Table 21-1 Impedance of a Series *RLi* Circuit

R (Ω)	L mH	E (V)	I (mA)	V_L (V)	$X_L = V_L/I$ (Ω)	$Z = E/I$ (Ω)	$Z = \sqrt{R^2 + X_L^2}$ (Ω)
470	25	5					
1000	50	5					

Table 21-2 Phase Angle of a Series *RL* Circuit

R (Ω)	d (Div)	ϕ (Deg)
1500		
560		

Table 21-3 Voltage Relationships of a Series *RL* Circuit

R (Ω)	E (V)	V_R (V)	V_L (V)	I (mA)	$X_L = 2\pi f L$ (Ω)	$\phi = \tan^{-1}(X_L/R)$ (Deg)	E $\sqrt{V_R^2 + V_L^2}$ (V)
560	5						
1500	5						

QUESTIONS

1. In your own words, describe the impedance relationship in a series *RL* circuit. Is this relationship proven in your experiment? Refer to your results.

2. Refer to your data in Table 21-1. How do the values of impedance calculated by Ohm's law and the impedance formula compare? Explain any discrepancies.

3. In your own words, explain the voltage relationship in a series *RL* circuit. Is this relationship proven in your experiment? Refer to your results.

4. How do the phase angles determined from oscilloscope measurements compare to the values determined from the X_L and R calculations. Explain any discrepancies.

____

____

____

____

____

____

5. An *RL* circuit has a resistance of 150-Ω and an inductance of 50-mH. At what frequency will its impedance be 200-Ω?

____

____

____

____

____

____

EXPERIMENT

22

THE SERIES *RC* CIRCUIT

OBJECTIVES

1. To verify the expression for total impedance of a series *RC* circuit.

$$Z=\sqrt{R^2+X_C^2}$$

2. To determine the phase angle between voltage and current in a series *RC* circuit.
3. To verify the voltage relationship of a series *RC* circuit.

$$E=\sqrt{V_R^2+V_C^2}$$

DISCUSSION

Series *RC* Circuit

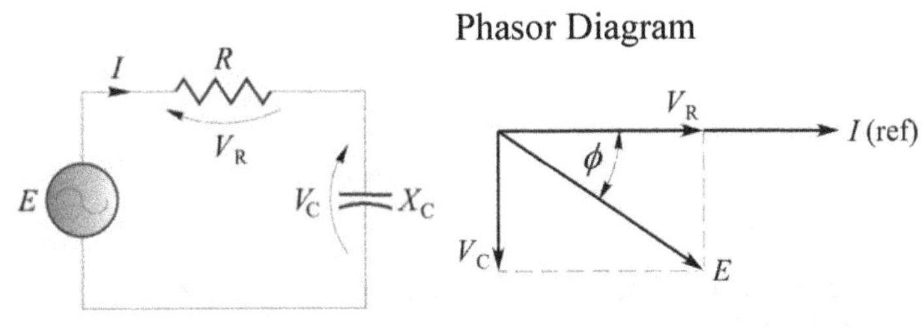

Current now *leads* the voltage by phase angle ϕ, which is between 0° and 90°.

Kirchhoff's voltage law is satisfied by vector addition of the voltages. ⇨ $|E|=\sqrt{V_R^2+V_C^2}$

The total opposition to AC current flow is again called the **impedance**, which has symbol **Z**.

$$Z=\sqrt{R^2+X_C^2} \text{ and } \phi=\tan^{-1}\left(\frac{V_C}{V_R}\right)$$

Impedance Triangle

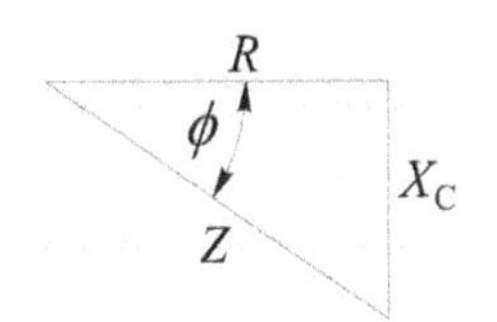

Example: A 10-μF capacitor is connected in series with a 200-Ω resistor across a 500-V, 60-Hz source. Calculate impedance, current, and phase angle.

$$X_C = \frac{1}{2\pi f C} = \frac{1}{377\,\text{rad/s} \times 10 \times 10^{-6}\,F} = 265.3\,\Omega$$

$$\Rightarrow\ Z = \sqrt{R^2 + X_C^2} = \sqrt{200^2 + 265.3^2} = \mathbf{332.2\,\Omega} \ \Rightarrow\ I = \frac{E}{Z} = \frac{500\,\text{V}}{332.2\,\Omega} = \mathbf{1.505\,A}$$

$$\phi = \tan^{-1}\left(\frac{X_L}{R}\right) = \tan^{-1}\left(\frac{265.3}{200}\right) = \mathbf{53°}$$

EQUIPMENT

Power Supplies:
- Function generator

Instruments:
- DMM
- Oscilloscope

Resistors: (5%, ½ W or higher)
- One each of 1-kΩ and 2.2-kΩ

Capacitors:
- One each of 0.047-μF and 0.22-μF; all rated 100 V

Others:
- Protoboard
- Hook-up Wire

PROCEDURE

PART A: Impedance of a Series *RC* Circuit

1. Connect the circuit shown in Figure 22-1. Set the function generator to a frequency of 2 kHz and increase the output to 5 V RMS (measured by DMM).

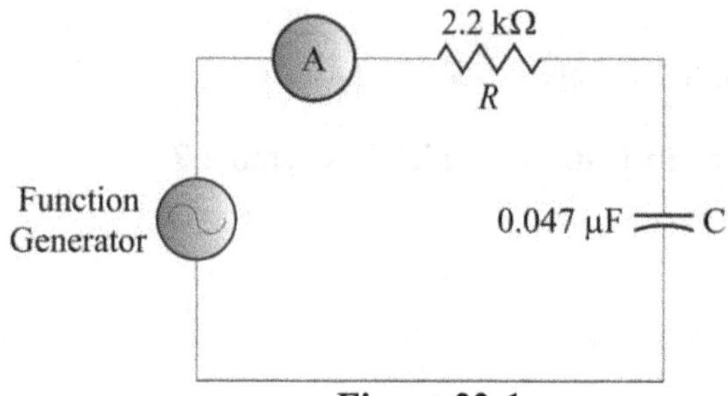

Figure 22-1

2. Measure current flowing in the circuit and record in Table 22-1. Measure the voltage across the capacitor and record in Table 22-1.

3. Calculate X_C from the measured values of V_C and I. Calculate total impedance from E/I. Record the values in Table 22-1.

4. Calculate the capacitive reactance of the capacitor from the formula $X_C = \frac{1}{2\pi f C}$ and record this value in Table 22-1. Determine the *calculated* value of impedance from the impedance formula and record this in Table 22-1.

5. Replace the resistor with a 1.0-kΩ resistor and the capacitor with a 0.10-μF capacitor. Repeat steps 2 to 5.

PART B: Phase and Voltage Relationships of a Series *RC* Circuit

6. Connect the circuit shown in Figure 22-2.

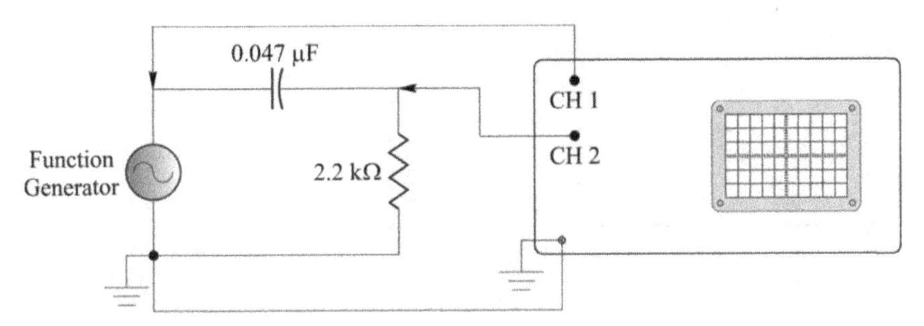

Figure 22-2

7. Switch **on** the oscilloscope. Set the trigger selector switch on the oscilloscope to CH 1. Adjust the horizontal and vertical controls to obtain a centered trace.

8. Power up the function generator and set the frequency to 2 kHz and increase the output to 5 V_{RMS} (measured by DMM).

9. Set the mode switch to CH 1. The sine wave on the oscilloscope is the reference wave and is the applied voltage in the circuit. Using the volts/div controls, adjust the height of the sine wave to about six divisions peak-to-peak. Using the sec (time)/div controls and the variable sec (time)/div, adjust the sine wave so that one cycle occupies exactly ten divisions. Each division will correspond to $360/10 = 36°$.

10. Set the mode switch to CH 2. Adjust the vertical control to obtain a centered trace. This sine wave is the voltage across the resistor and since resistor voltage is proportional to current, it also represents the current wave in this circuit. Using the volts/div controls, adjust the height of the sine wave to about four divisions peak-to-peak.

11. Set the mode switch to dual and measure the lead distance (d) between corresponding zero points on the voltage and current waves as shown in Figure 22-3.

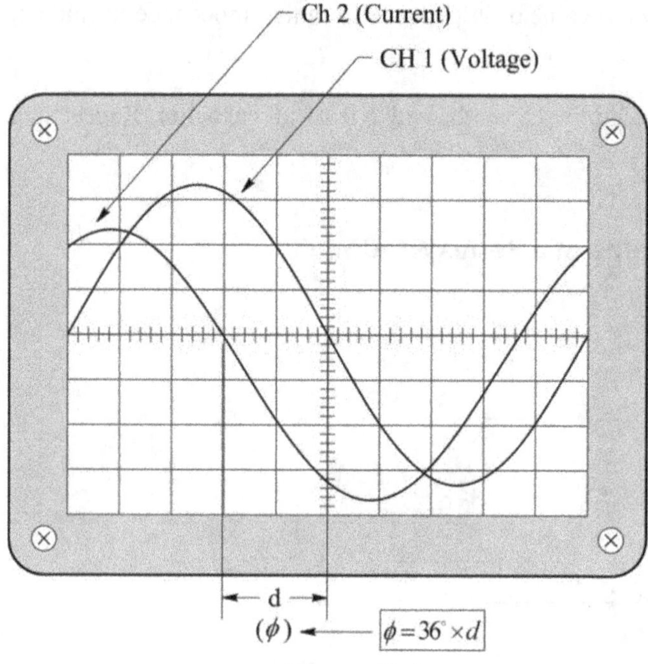

Figure 22-3

12. Determine the phase angle in degrees and record in Table 22-2.

13. Replace the 2.2-kΩ resistor with the 1-kΩ resistor and repeat steps 9 to 12.

14. Disconnect the oscilloscope from the circuit. Check the applied voltage to verify that it is still 5 V_{RMS} and adjust if necessary. Using the DMM, measure the voltages across the capacitor and inductor and record in Table 22-3.

15. Calculate the current in the circuit using Ohm's law from the measured values of V_R and R. Calculate the capacitive reactance, X_C, from the equation $X_C = \frac{1}{2\pi f C}$. Record these values in Table 22-3.

16. Calculate the phase angle of the circuit from $\phi = \tan^{-1}\left(\frac{X_C}{R}\right)$. Record this value in Table 22-3.

17. Using the vector relationship, calculate the applied voltage from the measured values of V_C and V_R.

$$E = \sqrt{V_R^2 + V_C^2}$$

18. Repeat steps 14 to 17 with the 2.2-kΩ resistor.

RESULTS

Table 22-1 Impedance of a Series *RC* circuit

R (Ω)	*C* (μF)	*E* (V)	*I* (mA)	V_R (V)	$X_C = V_C/I$ (Ω)	$Z = E/I$ (Ω)	Z $\sqrt{R^2 + X_C^2}$ (Ω)
2200	0.047	75					
1000	0.1	75					

Table 22-2 Phase Angle of a Series *RC* Circuit

R (Ω)	d (Div)	ϕ (Deg)
2200		
1000		

Table 22-3 Voltage Relationships of a Series *RC* Circuit

R (Ω)	*E* (V)	V_R (V)	V_C (V)	*I* (mA)	$X_C = 1/2\pi f C$ (Ω)	$\phi = \tan^{-1}(X_C/R)$ (Deg)	E $\sqrt{V_R^2 + V_C^2}$ (V)
1000	5						
2200	5						

QUESTIONS

1. In your own words, describe the impedance relationship in a series *RC* circuit. Is this relationship proven in your experiment? Refer to your results.

2. Refer to your data in Table 22-1. How do the values of impedance calculated by Ohm's law and the impedance formula compare? Explain any discrepancies.

3. How do the phase angles determined from the oscilloscope measurements compare to the values determined from the X_C and R calculations? Explain any discrepancies.

4. The *RC* series circuit and the *RL* series circuit have similar impedance and voltage relationships. What is the major difference between the two circuits?

____

____

____

____

____

____

5. What capacitance in series with a 1.5-kΩ resistor will draw 50 mA from a 120-V, 60-Hz source?

____

____

____

____

____

____

EXPERIMENT

23

EFFECT OF FREQUENCY CHANGES ON A REACTIVE CIRCUIT

OBJECTIVES

1. To investigate the effect of varying frequency on a series *RL* circuit.
2. To investigate the effect of varying frequency on a series *RC* circuit.

DISCUSSION

Series *RL* Circuit

The impedance of a series *RL* circuit is given by $Z = \sqrt{R^2 + X_L^{\ 2}}$.

Since $X_L = 2\pi\ f\ L$, the impedance will be affected by a change in frequency. For $f = 0$ Hz (DC), X_L will be zero, and the impedance will equal the resistance in the circuit. Current will be maximum. As frequency increases, X_L will increase and Z will also increase.

Current $I = \dfrac{E}{Z}$ will decrease as the frequency and therefore impedance increases.

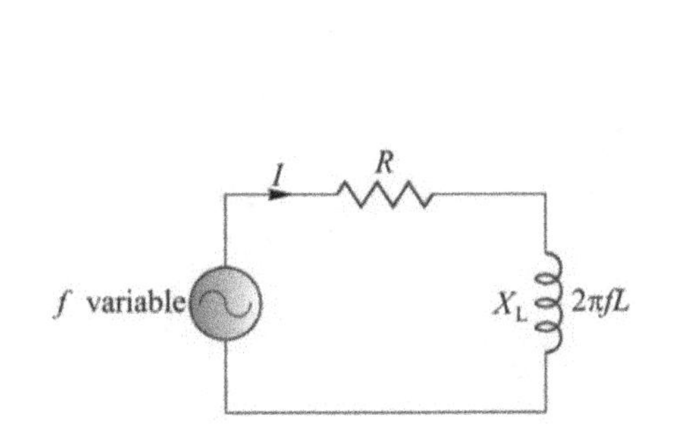

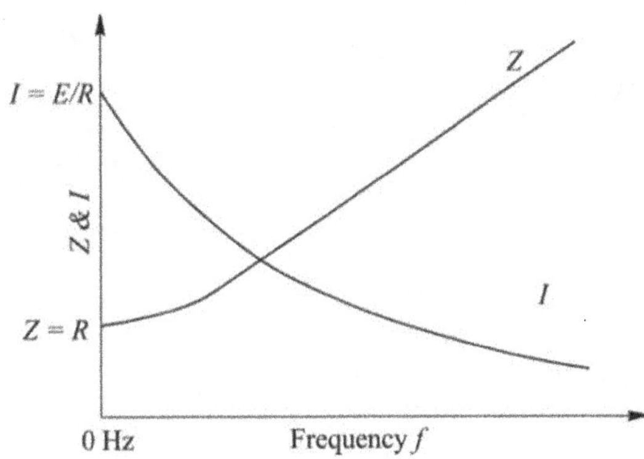

Series *RC* Circuit

The Impedance of a series *RC* circuit is given by $Z = \sqrt{R^2 + X_C^{\ 2}}$.

In this case $X_C = \dfrac{1}{2\pi f C}$ and will decrease as frequency increases. For $f = 0$ Hz (DC), X_C will be infinite, and so will Z (open circuit). No current will flow. As the frequency increases, X_C will decrease, therefore Z will decrease and current will increase. For high frequencies, X_C becomes very small and the total impedance approaches the value of resistance.

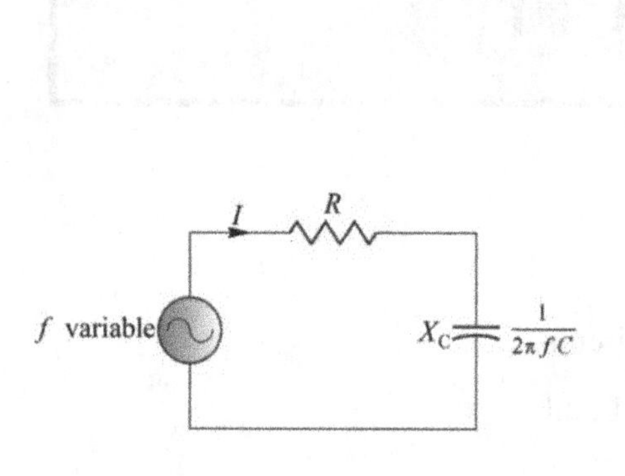

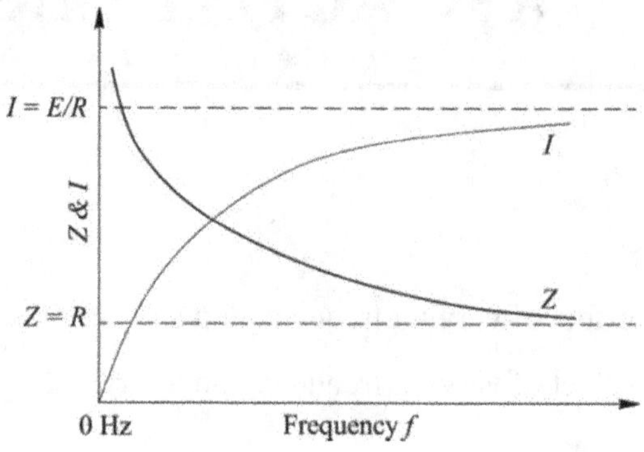

EQUIPMENT

Power Supplies:

- Function generator

Instruments:

- DMM

Capacitors:

- 0.1-μF, 100-V, DC

Inductor:

- 25-mH

Resistors: (5%, ½ W or higher)

- 820-Ω

Others:

- Breadboard
- Hook-up Wire

PROCEDURE

PART A: Frequency Response of a Series *RL* Circuit

1. Connect the circuit shown in Figure 23-1.

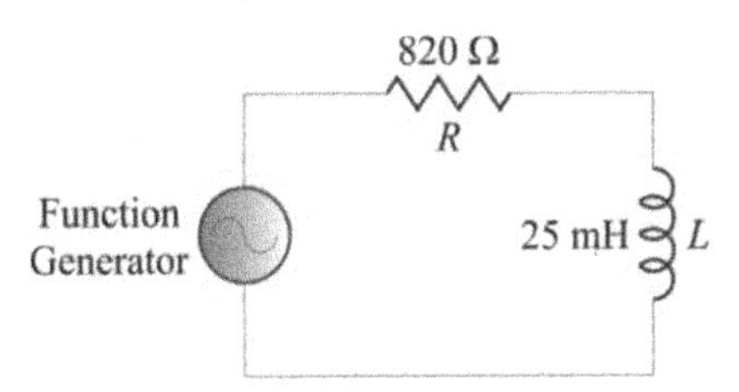

Figure 23-1

2. Set the function generator to its lowest frequency and output voltage. Switch **on** the function generator and set the frequency to 1 kHz. Increase the output voltage across the *RL* circuit to 5 V_{RMS}. This voltage must be maintained throughout this experiment.

3. Measure the voltages across the resistor and inductor and record in Table 23-1.

4. Increase the frequency to the next value shown in Table 23-1. Check the applied voltage and adjust to 5 V_{RMS} if necessary. Measure the voltages across the resistor and inductor and record in Table 23-1.

5. Repeat step 4 for each of the frequencies listed in Table 23-1.

6. *Calculate* the circuit current for each frequency from $I = \frac{V_R}{R}$ and record in Table 23-1.

7. *Calculate* the circuit impedance for each frequency from $Z = \frac{E}{I}$ and record these values in Table 23-1.

PART B: Frequency Response of a Series *RC* Circuit

8. Connect the circuit shown in Figure 23-2.

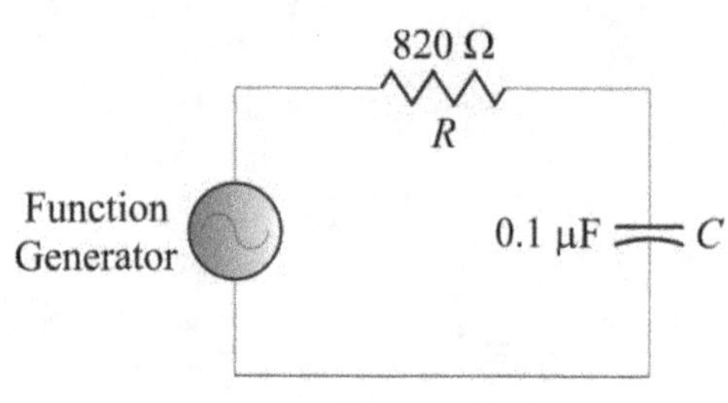

Figure 23-2

9. Set the function generator to its lowest frequency and output voltage. Switch **on** the function generator and set the frequency to 1 kHz. Increase the output voltage across the *RC* circuit to 5 V RMS. This voltage must be maintained throughout this experiment.

10. Measure the voltages across the resistor and capacitor and record in Table 23-2.

11. Increase the frequency to the next value shown in Table 23-2. Check the applied voltage and adjust to 5 V_{RMS} if necessary. Measure the voltages across the resistor and capacitor and record in Table 23-2.

12. Repeat step 11 for each of the frequencies listed in Table 23-2.

13. *Calculate* the circuit current for each frequency from $I = \frac{V_R}{R}$ and record in Table 23-2.

14. *Calculate* the circuit impedance for each frequency from $Z = \frac{E}{I}$ and record these values in Table 23-2.

RESULTS

Table 23-1 Frequency Response of a Series *RL* Circuit

f (kHz)	E (V)	V_R (V)	V_L (V)	$I = V_R/R$ (mA)	Z E/I (Ω)
1	5				
2	5				
3	5				
4	5				
5	5				
6	5				
7	5				
8	5				
9	5				
10	5				

Table 23-2 Frequency Response of a Series *RC* Circuit

f (kHz)	E (V)	V_R (V)	V_C (V)	$I = V_R/R$ (mA)	$Z = E/I$ (Ω)
1	5				
2	5				
3	5				
4	5				
5	5				
6	5				
7	5				
8	5				
9	5				
10	5				

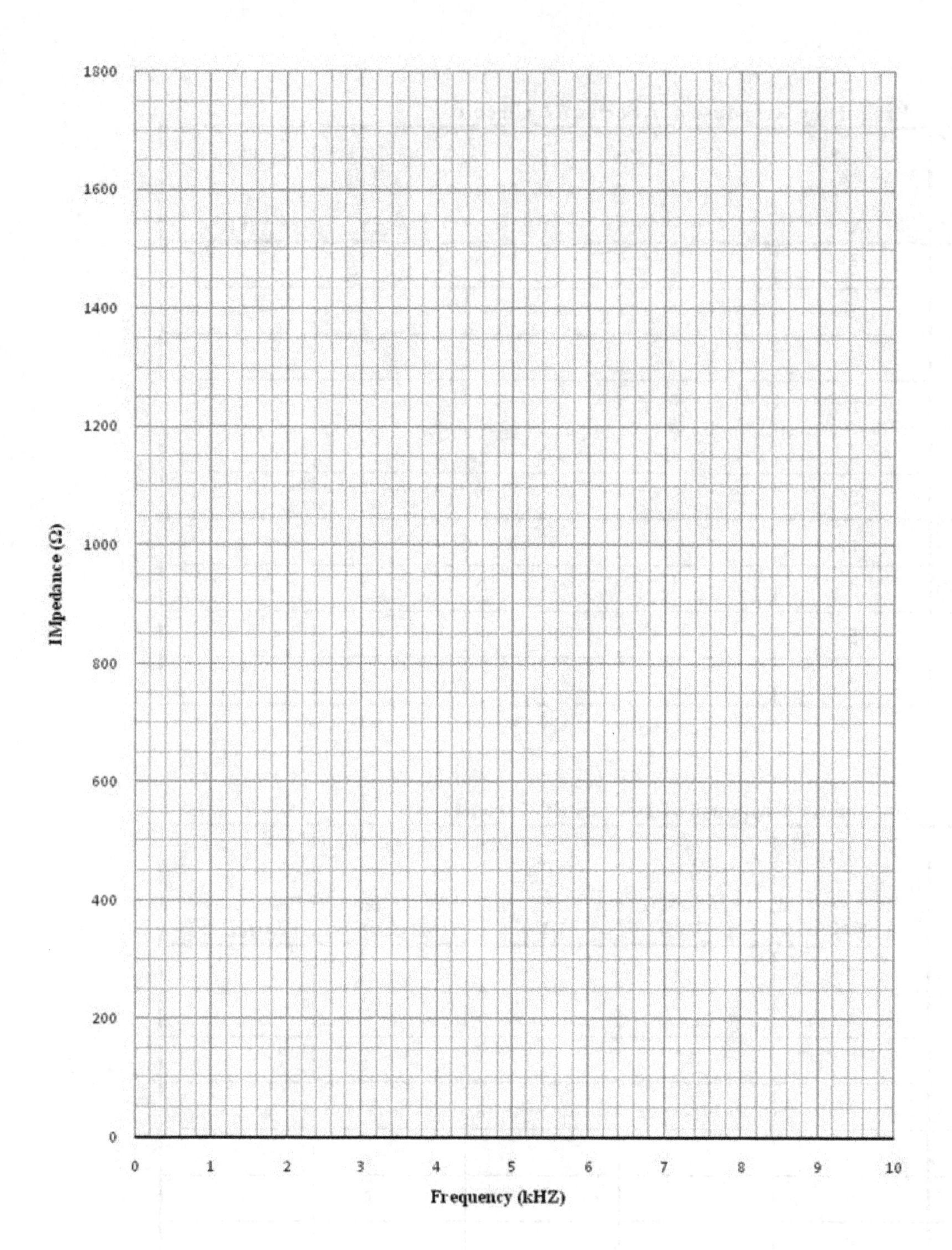
1800
1600
1400
1200
1000
800
600
400
200
0
IMpedance (Ω)
0
1
2
3
4
5
6
7
8
9
10
Frequency (kHZ)

QUESTIONS

1. Referring to your results, how does an increase in frequency affect the impedance of a series *RL* circuit? Explain your answer.

2. Referring to your results, how does an increase in frequency affect the impedance of a series *RC* circuit? Explain your answer.

3. Referring to your results, how are the voltages across the inductor and capacitor affected as the frequency increases? Explain why this happens.

4. On the sheet of graph paper provided, plot a graph of impedance vs. frequency for the *RL* and *RC* circuits. Plot frequency on the horizontal axis. From your graphs, determine the frequency at which the impedances of each circuit are equal.

5. For the *RL* and *RC* circuits used in this lab, calculate the frequency at which both circuits have the same impedance. How does this compare to the value obtained from the graphs?

EXPERIMENT

24

THE SERIES *RLC* CIRCUIT

OBJECTIVES

1. To verify experimentally that the impedance of a series *RLC* circuit is given by the following expression:

$$Z=\sqrt{R^2+X_T^2} \quad \text{where } X_T = X_L - X_C \text{ (net reactance)}.$$

2. To verify the voltage relationship for a series *RLC* circuit.

$$E=\sqrt{V_R^2+(V_L-V_C)^2}$$

DISCUSSION

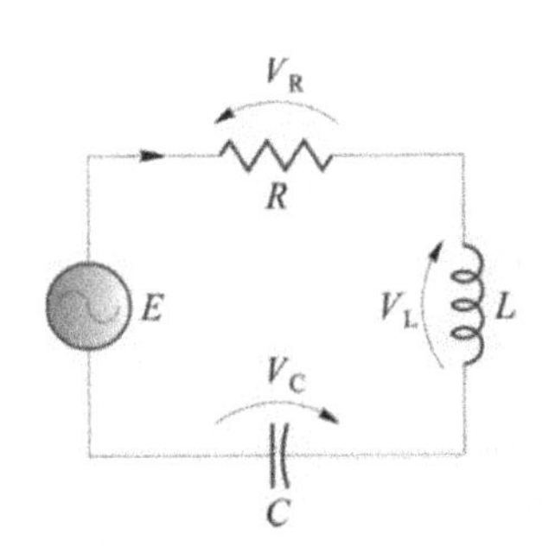

Phasor Diagrams

The current may *lag* or *lead* depending on the magnitudes of X_L and X_C.

(a) $X_L > X_C$ Lagging Phase Angle

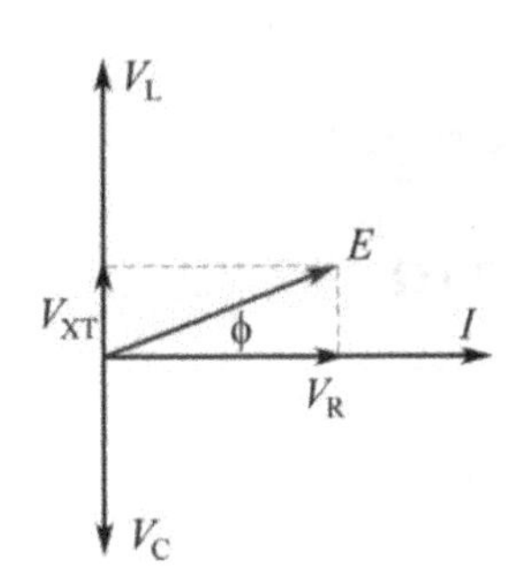

(b) $X_C > X_L$ Leading Phase Angle

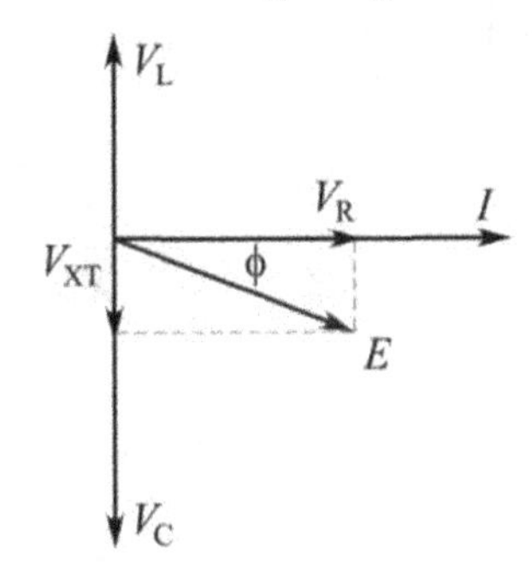

The equation for voltage, E, now becomes ⇨ $\Rightarrow E = \sqrt{V_R^2 + (V_L - V_C)^2}$

where $V_L - V_C$ is the *difference* between the capacitor and inductor voltages.

The equation for the circuit phase angle is ⇨ $\sqrt{\boldsymbol{V_R^2 + (V_L - V_C)^2}}$

The total opposition to AC current flow (impedance) is now a combination of resistance, inductive reactance, and capacitive reactance.

$$Z = \sqrt{R^2 + X_T^2} \quad where \quad X_T = X_L - X_C = \boldsymbol{net\ reactance}$$

Impedance Triangles

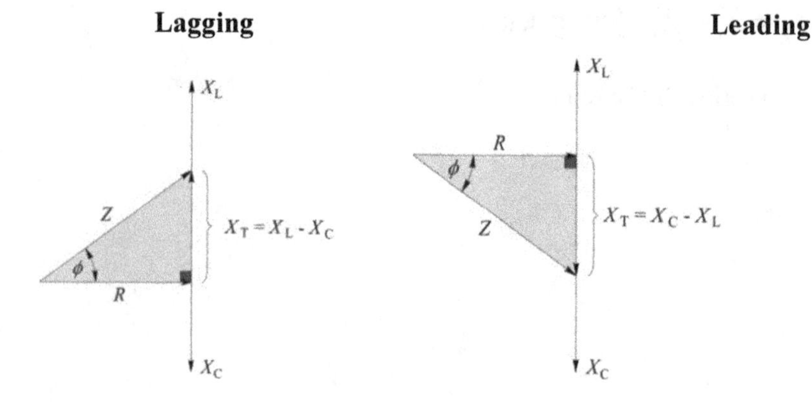

The phase angle, ϕ, can also be determined from the impedance Δ ⇨ $\phi = \tan^{-1}\left(\frac{X_T}{R}\right)$

Example: A 200V, 60Hz, power supply is connected to an R-L-C series circuit consisting of a 30Ω resistor, a 0.8H inductor, and a 10μF capacitor. Calculate the following:

(a) impedance
(b) current
(c) phase angle
(d) voltage across each component

(a) $X_L = 2\pi f L = 377\,\text{rad/s} \times 0.8\,\text{H} = 301.6\,\Omega$

$$X_C = \frac{1}{2\pi f C} = \frac{1}{377\,\text{rad/s} \times 10 \times 10^{-6}\,\text{F}} = 265.3\,\Omega \quad \therefore \text{ The circuit is inductive since } X_L > X_C$$

$$X_T = X_L - X_C = 36.3\,\Omega \;⇨\; Z = \sqrt{R^2 + X_T^2} = \sqrt{30^2 + 36.3^2} = \sqrt{2217.7} = \mathbf{47.1\,\Omega}$$

(b) $I = \frac{E}{Z} = \frac{200\,\text{V}}{47.1\,\text{A}} = \mathbf{4.25\,A1}$ (c) $\phi = \tan^{-1}\left(\frac{X_T}{R}\right) = \tan^{-1}\left(\frac{36.3}{30}\right) = \mathbf{50.4°}$ lagging

(d) $V_R = I \times R = 4.25\,\text{A} \times 30\,\Omega = \mathbf{127.4\,V}$

$V_L = I \times X_L = 4.25\,\text{A} \times 301.6\,\Omega = \mathbf{1280.7\,V}$

$V_C = I \times X_C = 4.25\,\text{A} \times 265.3\,\Omega = \mathbf{1126.5\,V}$

Check: $E = \sqrt{V_R^2 + (V_L - V_C)^2} = \sqrt{127.4^2 + 154.2^2} = \mathbf{200\,V}$

EQUIPMENT

Power Supplies:
- Isolation transformer
- Function generator
- Oscilloscope

Instruments:
- DMM

Resistors: (5%, ½ W or higher)
- 560-Ω

Capacitors:
- One each of 0.047-μF and 1.0-μF; all rated at 100-V

Inductor:
- 25-mH

Other:
- Protoboard
- Hook-up Wire

PROCEDURE

PART A: Impedance and Voltage Relationships of a Series *RLC* Circuit

1. Connect the circuit shown in Figure 24-1.

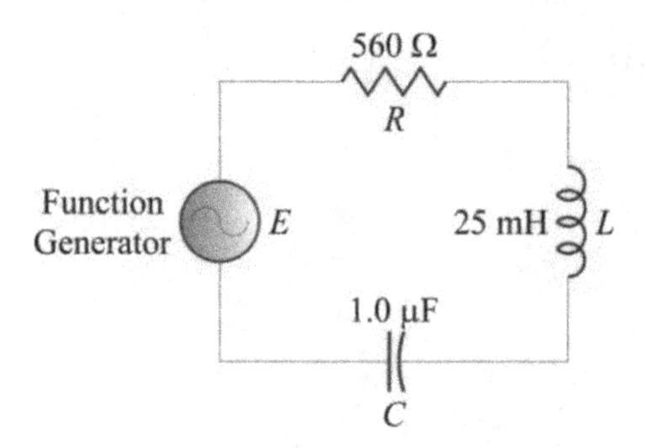

Figure 24-1

2. Switch **on** the function generator and set the frequency to 3 kHz and adjust the output voltage (E) to 5 V_{RMS}.

3. Measure the voltages across the resistor, inductor, and capacitor. Record your values in Table 24-1.

4. Replace the capacitor with a 0.047-μF capacitor and repeat step 3. Make sure the source voltage remains at 5 V.

5. Increase the frequency to 4.5 kHz and repeat step 3. Make sure the source voltage remains at 5 V_{RMS}.

6. For each step, calculate the circuit current from $I = {}^{V_R}\!/_{R}$. Calculate circuit impedance from $Z = {}^{E}\!/_{I}$. Calculate X_L and X_C from their respective formulas and calculate theoretical impedance from $Z = \sqrt{R^2 + X_T^2}$.

   Calculate total voltage from the vector equation $E = \sqrt{V_R^2 + V_{X_T}^2}$ $where\ V_{X_T} = V_L - V_C$. Enter all values in Table 24-1.

PART B: Phase Angle of a Series *RLC* Circuit

7. Connect the circuit shown in Figure 24-2.

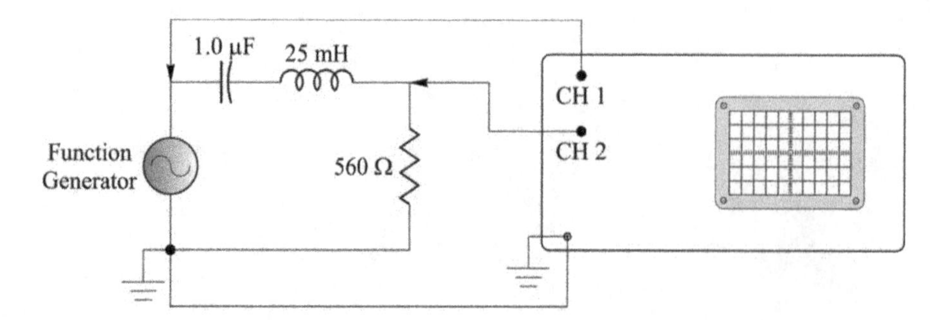

Figure 24-2

8. Turn on the function generator, set the frequency to 3 kHz, and increase the output to approximately mid-range.

9. Using the techniques from previous experiments, measure the phase shift between CH 1 (voltage) and CH 2 (current) in divisions and convert to degrees. The angle may be *lagging* or *leading*, so make sure to measure between the correct zero points on the waveforms. Record the values in Table 24-2.

10. Replace the capacitor with a 0.047-μF capacitor and repeat step 9.

11. Calculate the theoretical phase angle for each circuit from $\phi = \tan^{-1}\left(\frac{X_T}{R}\right)$ and enter these values in Table 24-2.

RESULTS

Table 24-1 Impedance and Voltage of a Series *RLC* circuit

C (μF)	f (kHz)	V_R (V)	V_C (V)	V_L (V)	$I = V_R/R$ (Ω)	X_C (Ω)	X_L (Ω)	Z E/I (Ω)	Z $\sqrt{R^2 + X_T^2}$ (Ω)	E $\sqrt{V_R^2 + V_X^2}$ (Ω)
1.0	3									
0.047	3									
0.047	4.5									

Table 24-2 Phase Angle of a Series *RLC* Circuit

C (μF)	d (Div)	ϕ (Deg)	Phase Shift (Lag/Lead)	ϕ (Calc)
1.0				
0.047				

QUESTIONS

1. In your own words, describe the impedance relationship in a series *RLC* circuit. Is this relationship proven in your experiment? Refer to your results and explain any discrepancies.

2. In your own words, explain the voltage relationship in a series *RLC* circuit. Is this relationship proven in your experiment? Refer to your results and explain any discrepancies.

3. In Table 24-1, for the 0.047-μF capacitor at 3 kHz, the voltage across the capacitor is greater than the source voltage. Explain how this is possible.

4. How do the measured values of phase angles in Table 24-2 compare with the calculated values? What did you notice about the phase shifts observed on the scope for the two circuits? Explain why this happened.

5. Under what conditions will the current in a series *RLC* circuit be at a maximum value? Did any of the circuits investigated approach this condition.

6. An inductive circuit of 50-Ω resistance and 0.08-H inductance is connected in series with a capacitor across a 200-V, 50-Hz supply. The current taken is 3.8 A leading. Find the value of the capacitor.

[illegible]

[illegible]

EXPERIMENT

25

RESONANCE OF A SERIES *RLC* CIRCUIT

OBJECTIVES

1. To determine experimentally the resonant frequency of a series *RLC* circuit.
2. To investigate experimentally the effect of varying frequency on the total impedance and current of a series *RLC* circuit.

DISCUSSION

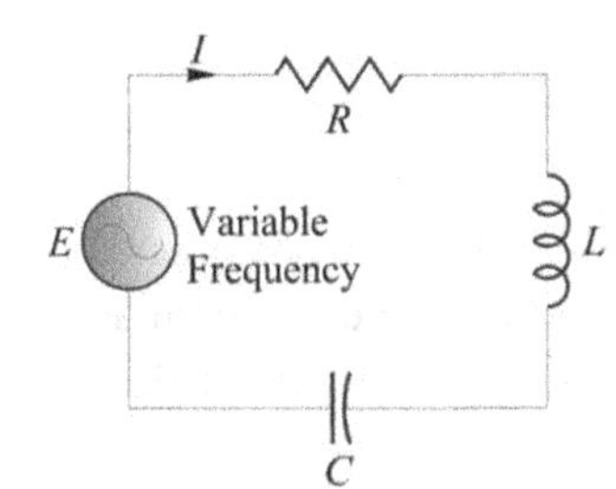

The total impedance, (Z), of this circuit is a combination of resistance, inductive reactance, and capacitive reactance.

$$Z = \sqrt{R^2 + X_T^2} \quad \Rightarrow X_T = X_L - X_C = (2\pi f L) - \left(\frac{1}{2\pi f C}\right)$$

This indicates that the impedance of this circuit is *frequency dependent* and will change as the source frequency is varied.

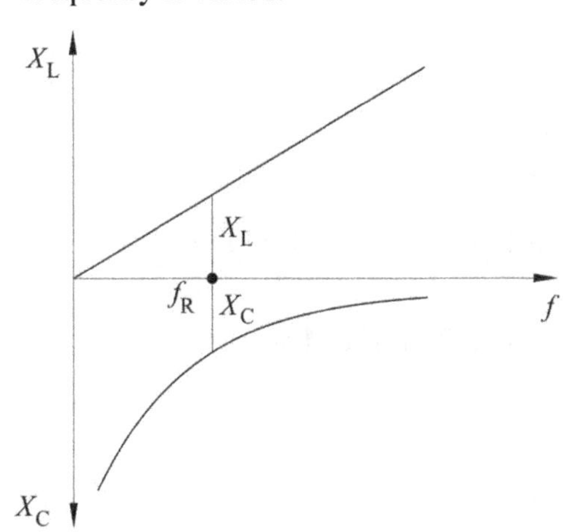

If frequency is started at a low value, X_C will be large and X_L will be small. Therefore X_T will be large, giving large total impedance. As frequency is increased, X_L increases and X_C decreases.

A frequency will be reached where $X_L - X_C$ and the reactances will cancel out, leaving no net reactance. The total impedance is now at its minimum and is simply the resistance of the circuit. This condition is called *resonance* and the frequency at which it occurs is called the *resonant frequency* (f_R).

If the frequency is increased beyond this point, X_L increases further and X_C decreases, there is again a net reactance, and the impedance will increase.

At resonance:

$$X_L = X_C$$

$$X_L = 2\pi f_R L \quad \text{and} \quad X_C = \frac{1}{2\pi f_R C} \qquad (f_R = \text{resonant frequency})$$

$$\Rightarrow\ 2\pi f_R L = \frac{1}{2\pi f_R C}\ \Rightarrow\ 4\pi^2 f_R^2 LC = 1\ \Rightarrow\ f_R^2 = \frac{1}{4\pi^2 LC}\ \Rightarrow\ f_R = \frac{1}{2\pi\sqrt{LC}}$$

Note: Resonant frequency is independent of resistance (R). However, resistance will determine the current that will flow at resonance.

At resonance $I = \dfrac{E}{R}$

Graph of Z vs. f

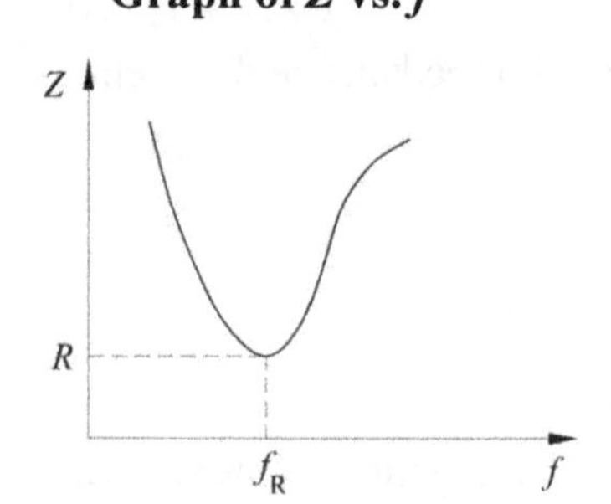

Graph of I vs. f

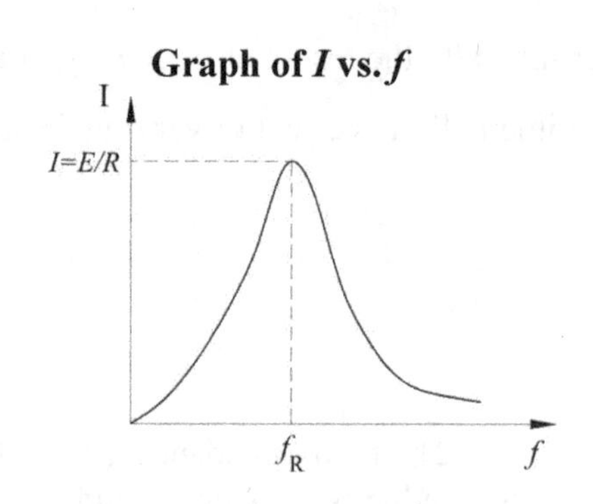

Example: A coil in a tuned circuit in a radio receiver has an inductance of 300 μH and a resistance of 15 Ω. What value of capacitance must be placed in series with the coil to be resonant at 840 kHz? If the signal strength of the radio signal is 250 μV, how much current will be flowing at resonance?

$$f_R = \frac{1}{2\pi\sqrt{LC}}\ \Rightarrow\ C = \frac{1}{4\pi^2 L f_R^2} = \frac{1}{4\pi^2 \times 300\times10^{-6} \times (840\times10^3)^2} = 1.2\times10^{-10}\ \text{F} = \mathbf{120\,pF}$$

$$I_{\text{res}} = \frac{E}{R} = \frac{250\,\mu V}{15\,\Omega} = \mathbf{16.7\,\mu A}$$

EQUIPMENT

Power Supplies:
- Function generator

Instruments:
- DMM
- Oscilloscope

Resistors:
- One each of 100-Ω and 1.0-kΩ

Inductors:
- One each of 25-mH and 50 mH

Capacitors:
- One each of 0.001-μF, 0.0047-μF, and 0.047-μF; all rated 100-V

Other:
- Protoboard
- Hook-up Wire

PROCEDURE

PART A: Resonant Frequency of a Series *RLC* Circuit

1. Connect the circuit shown in Figure 25-1.

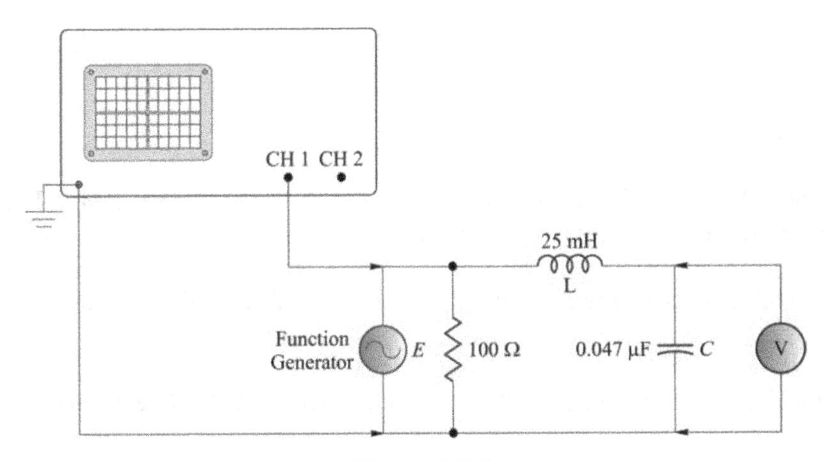

Figure 25-1

2. Switch on the function generator. Adjust the frequency of the function generator to 5 kHz and the output to 10-V peak-to-peak as measured on the oscilloscope. The 100-Ω resistor is *not* part of the test circuit and is used to stabilize the output voltage as the impedance of the circuit changes with frequency.

3. Observe the voltage across the capacitor as the frequency of the function generator is varied above and below 5000 Hz.

4. Determine the frequency at which the capacitor voltage is *maximum*. This is the resonant frequency, f_R, of the circuit. Record this value in Table 25-1.

5. Calculate the resonant frequency of the circuit from the rated values of capacitance and inductance. Record the value in Table 25-1.

6. Replace the 0.047-μF capacitor with a 0.0047-μf capacitor. Repeat steps 4 and 5, starting at 15 kHz.

7. Replace the 0.0047-μF capacitor with a 0.001-μf capacitor and the inductor with a 50-mH inductor. Repeat steps 4 and 5 starting at 20 kHz.

PART B: Frequency Response of a Series *RLC* Circuit

8. Connect the circuit shown in Figure 25-2.

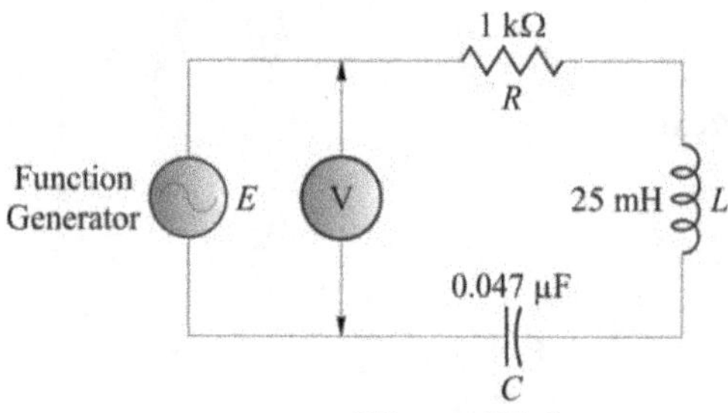

Figure 25-2

9. Switch **on** the function generator. Adjust the frequency to 1000 Hz and the output to 5 V RMS. **This voltage must be maintained for each step of this experiment.**

10. Using the DMM, measure the voltages across the resistor, capacitor, and inductor. Record your results in Table 25-2.

11. Increase the frequency to 2000 Hz, check the source voltage and adjust to 5V RMS if necessary. Repeat step 10.

12. Adjust the frequency to each of the values shown in Table 25-2 and repeat step 10. Make sure that the source voltage remains at 5 V_{RMS} for each step.

13. Connect the oscilloscope across the inductor and capacitor in series as shown in Figure 25-3. Adjust the sec/div to display approximately four cycles of the waveform.

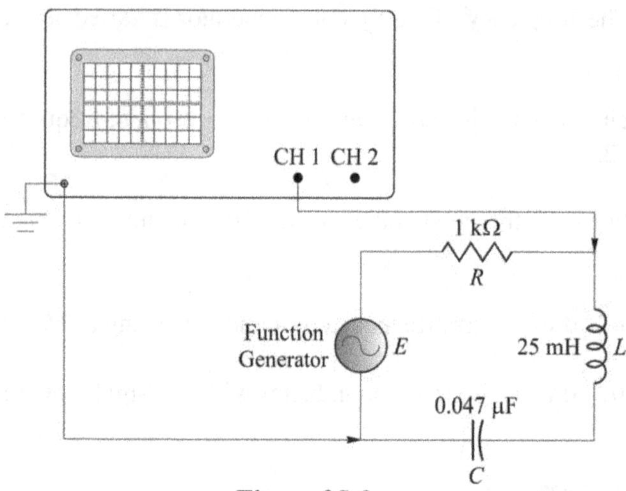

Figure 25-3

14. Vary the frequency until the amplitude of the waveform is at a minimum. This is the resonant frequency. Record this frequency under f_R in Table 25-2.

15. For each frequency in Table 25-2, calculate the current in the circuit from the measured value of V_R and the rated resistor value. Calculate the capacitive and inductive reactance from the measured values of V_L and V_C and the calculated current. Calculate the total circuit impedance from the source voltage and the calculated current.

16. Plot a graph of Z vs. f using the data from Table 25-2. Plot frequency on the horizontal axis. Indicate the resonant frequency on your graph.

RESULTS

Table 25-1 Resonant Frequency of a Series *RLC* Circuit

L (mH)	C (μF)	f_R (kHz) Measured From Circuit	Calculated $f_R = \frac{1}{2\pi\sqrt{LC}}$
25	0.047		
25	0.0047		
50	0.004		

Table 25-2 Frequency Response of a Series *RLC* Circuit

f (Hz)	E (V)	V_R (V)	V_L (V)	V_C (V)	I V_R/R (mA)	$X_L = V_L/I$ (Ω)	$X_C = V_C/I$ (Ω)	$Z = E/I$ (Ω)
1000	5							
2000	5							
3000	5							
4000	5							
5000	5							
6000	5							
7000	5							
8000	5							
10 000	5							
12 000	5							
15 000	5							
20 000	5							
f_R	5							

IMPEDANCE vs. FREQUENCY

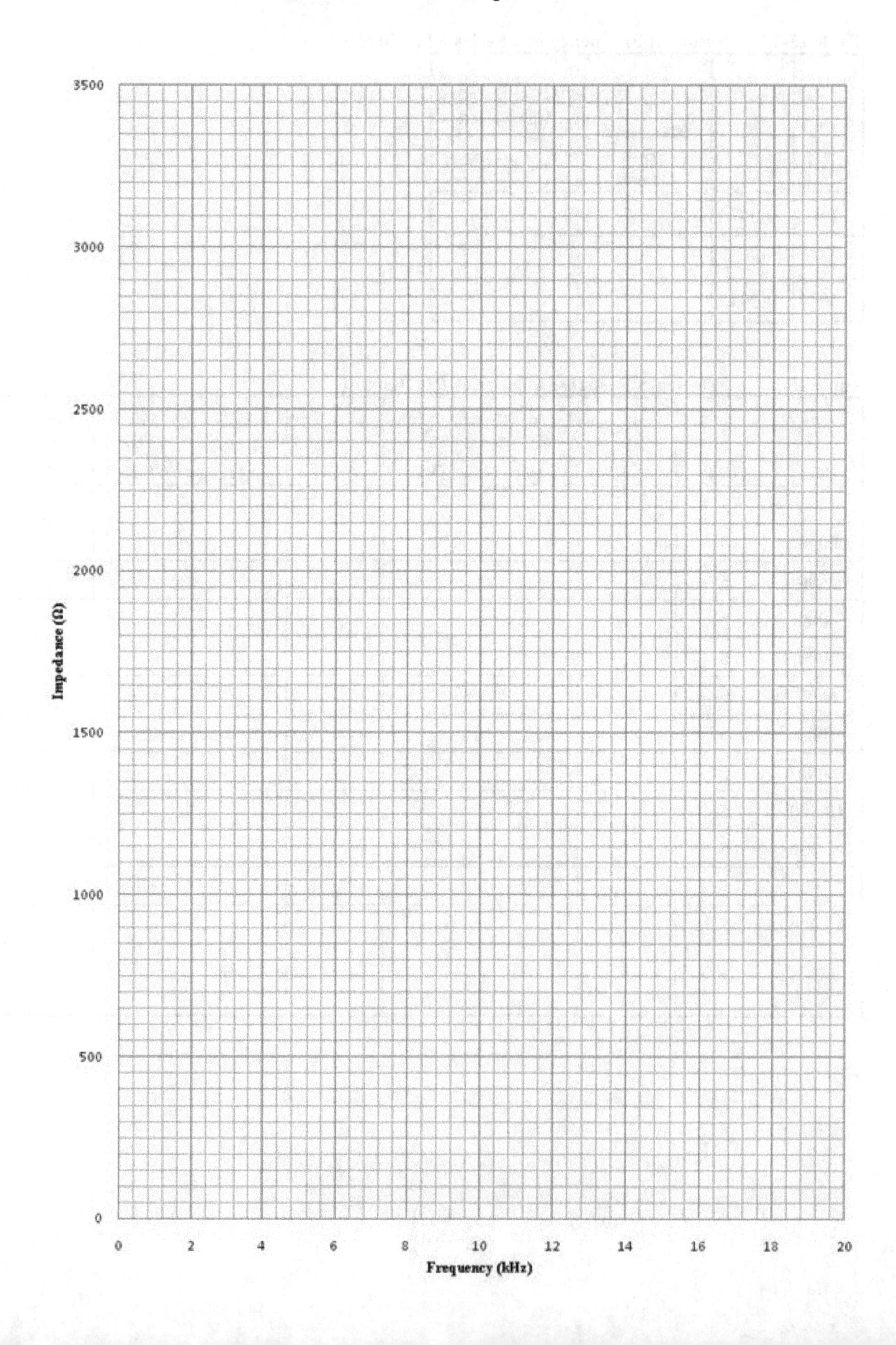

QUESTIONS

1. In your own words, describe the effect of frequency on the impedance of a series *RLC* circuit. Use your results to explain this effect.

2. Calculate the current flowing at resonance for the experimental circuit shown in Figure 25-2. How does this compare to the measured value? Explain any discrepancies.

3. Describe a practical application for a circuit displaying resonance effects.

4. From your experimental data in Table 25-2, what is the impedance of the circuit at resonance? How does this compare to the theoretical value? Explain.

5. A 1500-pF capacitor is used in a tuned circuit with an inductor having a resistance of 12 Ω. What is the value of the inductance if the resonant frequency is 291 kHz? To what frequency will the circuit tune if the capacitance is changed to 500 pF?

EXPERIMENT

26

QUALITY FACTOR AND BANDWIDTH OF A RESONANT CIRCUIT

OBJECTIVES

1. To investigate the frequency response of a series *RLC* circuit with varying L and C.
2. To determine the bandwidth from the frequency response curve.
3. To investigate the effect of the quality factor (Q) on bandwidth.

DISCUSSION

Selectivity of a Series Resonant Circuit

The selectivity of a tuned circuit is its ability to select (tune) one particular frequency and reject all others.

A perfectly selective circuit would have the characteristic shown in the first diagram

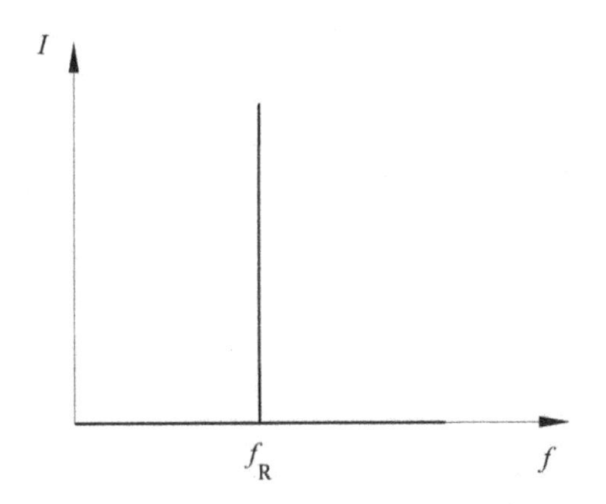

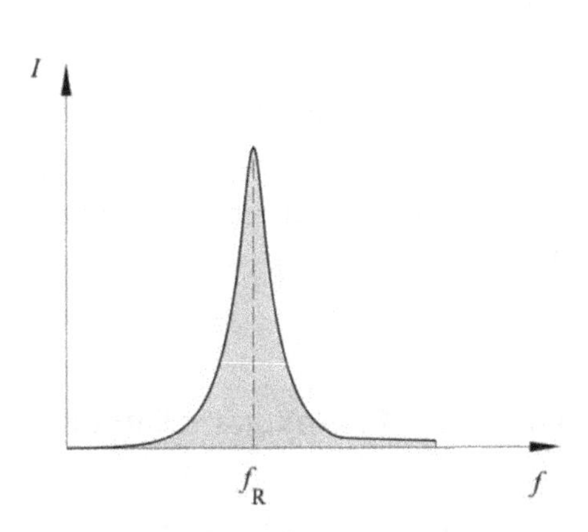

In a practical example, a selective circuit is one having a very sharp peak as shown in the second diagram.

Factors Influencing Selectivity

1. **Resistance**: Decreasing resistance increases the current that flows at resonance, giving a sharper peak.

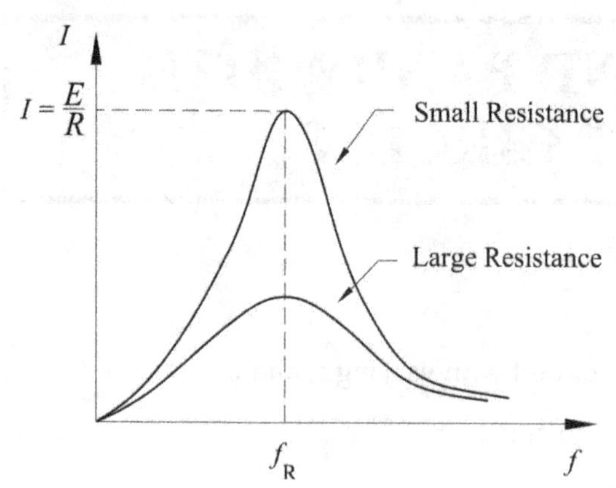

This effect is limited by the resistance of the inductor.

2. **The $\frac{L}{C}$ ratio**

Since $f_R = \dfrac{1}{2\pi\sqrt{LC}}$, therefore to maintain f_R as a constant, the product LC must remain constant and the $\frac{L}{C}$ ratio can only be changed under this condition.

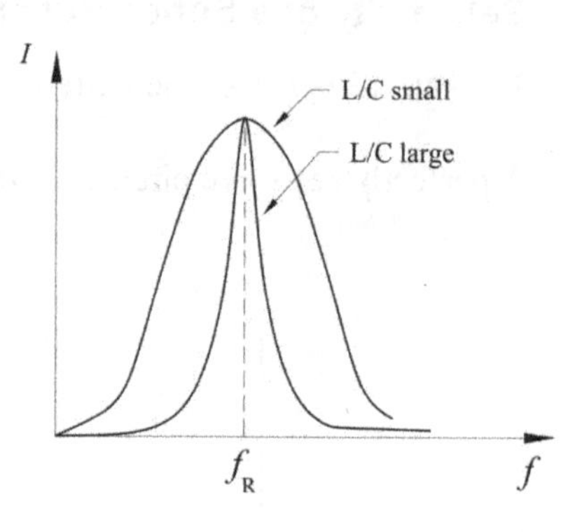

Increasing the $\frac{L}{C}$ ratio steepens the skirts of the curve, since the net reactance $X_T = (X_L - X_C)$ will be greater for off-resonance frequencies. This reduces the current for off-resonance frequencies, which gives a sharper curve.

Quality Factor (Q) of a Resonant Circuit

➢ The Q-factor of an RL circuit was defined as ⇨ $Q = \dfrac{X_L}{R} = \dfrac{2\pi f L}{R}$

But at resonance, $f = f_R = \dfrac{1}{2\pi\sqrt{LC}}$ ⇨ $Q = \dfrac{2\pi \times \left(\dfrac{1}{2\pi\sqrt{LC}}\right) \times L}{R}$

$$Q = \frac{L}{R\sqrt{LC}} = \frac{1}{R}\sqrt{\frac{L^2}{LC}} = \frac{1}{R}\sqrt{\frac{L}{C}}$$

Since Q is inversely proportional to resistance and directly proportional to $\frac{L}{C}$, it will be a good indication of the shape and selectivity of the tuned circuit. A higher Q means a steeper curve and therefore a more selective circuit.

Bandwidth

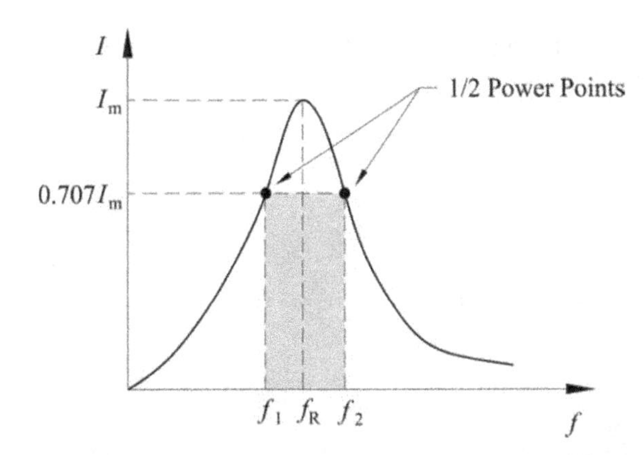

As shown from the resonant curve above, for frequencies close to the resonant frequency, the current is only slightly smaller than the current at resonance. Therefore, a practical circuit does not select a single frequency, but rather a band of frequencies between f_1 and f_2.

The cut-off frequencies f_1 and f_2 are the frequencies at which the current has dropped to 70.7% of its maximum value.

The bandwidth (Δf) is defined as the difference between the upper and lower frequencies f_1 and f_2.

$$\Delta f = f_2 - f_1$$

At resonance: $I = I_m \Rightarrow P_m = I_m^2 R$

At the cut-off points:

$$I = 0.707 I_m = \frac{1}{\sqrt{2}} I_m \Rightarrow P = I^2 R = \left(\frac{1}{\sqrt{2}} I_m\right)^2 R \Rightarrow P = \frac{1}{2} I_m^2 R$$

$$\boldsymbol{P = \frac{1}{2} P_m}$$

The power at the cut-off points is half of the maximum power ($\boldsymbol{P_m}$). Thus the upper and lower cut-off points are also called the **half-power points**.

The bandwidth of a circuit depends on the shape of the resonance curve. Steep sides mean a narrow bandwidth. Since the Q factor is an indicator of the steepness of the curve, it follows that bandwidth will be inversely proportional to Q. It can be shown that $\Delta f = \frac{f_R}{Q}$.

Assuming that the curve is symmetrical with respect to the resonant frequency, the cut-off frequencies can be found from the following equations:

$f_1 = f_R - \frac{\Delta f}{2}$ and $f_2 = f_R + \frac{\Delta f}{2}$

Example: Determine the bandwidth and cut-off frequencies of a series resonant circuit consisting of a 50-μH inductor having a 50-Ω resistance and a 200-pF capacitor.

$$f_R = \frac{1}{2\pi\sqrt{LC}} = \frac{1}{2\pi\sqrt{(50\times10^{-6})\times(200\times10^{-12})}} = 1590\,\text{kHz}$$

$$Q = \frac{1}{R}\sqrt{\frac{L}{C}} = \frac{1}{50}\sqrt{\frac{50\times10^{-6}}{200\times10^{-12}}} = 10 \;\Rightarrow\; \Delta f = \frac{f_R}{Q} = \frac{1590\,\text{kHz}}{10} = \mathbf{159\,kHz}$$

$$f_1 = f_R - \frac{\Delta f}{2} = 1590\,\text{kHz} - \frac{159\,\text{kHz}}{2} = \mathbf{1511\,kHz}$$

$$f_2 = f_R + \frac{\Delta f}{2} = 1590\,\text{kHz} + \frac{159\,\text{kHz}}{2} = \mathbf{1670\,kHz}$$

EQUIPMENT

Power Supplies:
- Function generator

Instruments:
- DMM

Resistors:
- 220-Ω

Capacitors:
- One each of 0.1-μF, 0.22-μF, and 0.47-μF; all rated at 100-V

Inductors:
- One each of 25-mH, 50-mH, and 100-mH

Other:
- Protoboard
- Hook-up Wire

PROCEDURE

1. Measure the resistance of each of the three inductors and record the values in Table 26-1.

2. Connect the circuit shown in Figure 26-1.

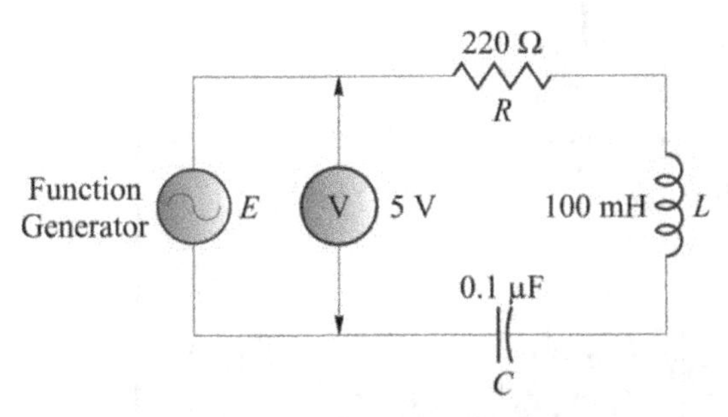

Figure 26-1

3. Switch **on** the function generator. Adjust the frequency to 600 Hz and the output to 5 V RMS. **This voltage must be maintained for each step of this experiment.**

4. Using the DMM, measure the voltage across the resistor. Record your results in Table 26-2.

5. Increase the frequency to 800 Hz. Check the source voltage and adjust to 5 V_{RMS} if necessary. Repeat step 3.

6. Adjust the frequency to each of the values shown in Table 26-3 and repeat step 4. **Make sure that the source voltage remains at 5 V for each step.**

7. Calculate the current flowing for each frequency from V_R/R and record the values in Table 26-2.

8. Change the capacitor to 0.22-μF and inductor to 50-mH. Repeat steps 3 to 7.

9. Change the capacitor to 0.47-μF and inductor to 25-mH. Repeat steps 3 to 7.

10. Calculate the *L*/*C* ratio for each combination and enter these values in Table 26-3.

11. On the same set of axes, draw graphs of *I* vs. *f* for each combination. Plot *f* on the x-axis.

12. On your graphs, determine the resonant frequency for each combination. Indicate the half-power points and use these to determine the bandwidth. Calculate *Q* for each combination from the values of f_R and Δf. Enter these values in Table 26-3.

13. Using the rated values of *L* and *C* and including the resistance of each inductor, calculate the theoretical values of f_R, Δf, and *Q*.

RESULTS

Table 26-1 Inductor Measured Resistance

Inductor (mH) →	25	50	100
Resistance (Ω)			

Table 26-2 Frequency Response of the *RLC* Circuit

		Step 4		Step 6		Step 8	
f (Hz)	E (V)	V_R (V)	$I = V_R/R$ (mA)	V_R (V)	$I = V_R/R$ (mA)	V_R (V)	$I = V_R/R$ (mA)
600	5						
800	5						
1000	5						
1200	5						
1300	5						
1400	5						
1500	5						
1600	5						
1700	5						
1800	5						
2000	5						
2500	5						
3000	5						

Table 26-3 Quality Factor and Bandwidth of an *RLC* Series Resonant Circuit

			Calculated			Measured from Graphs		
L (mH)	C (μF)	$\frac{L}{C}, (\frac{H}{F})$	f_R	Q	Δf	f_R	Q	Δf
100	0.1							
50	0.22							
25	0.47							

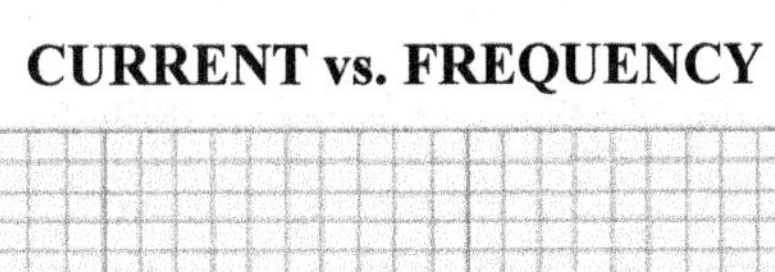
CURRENT vs. FREQUENCY

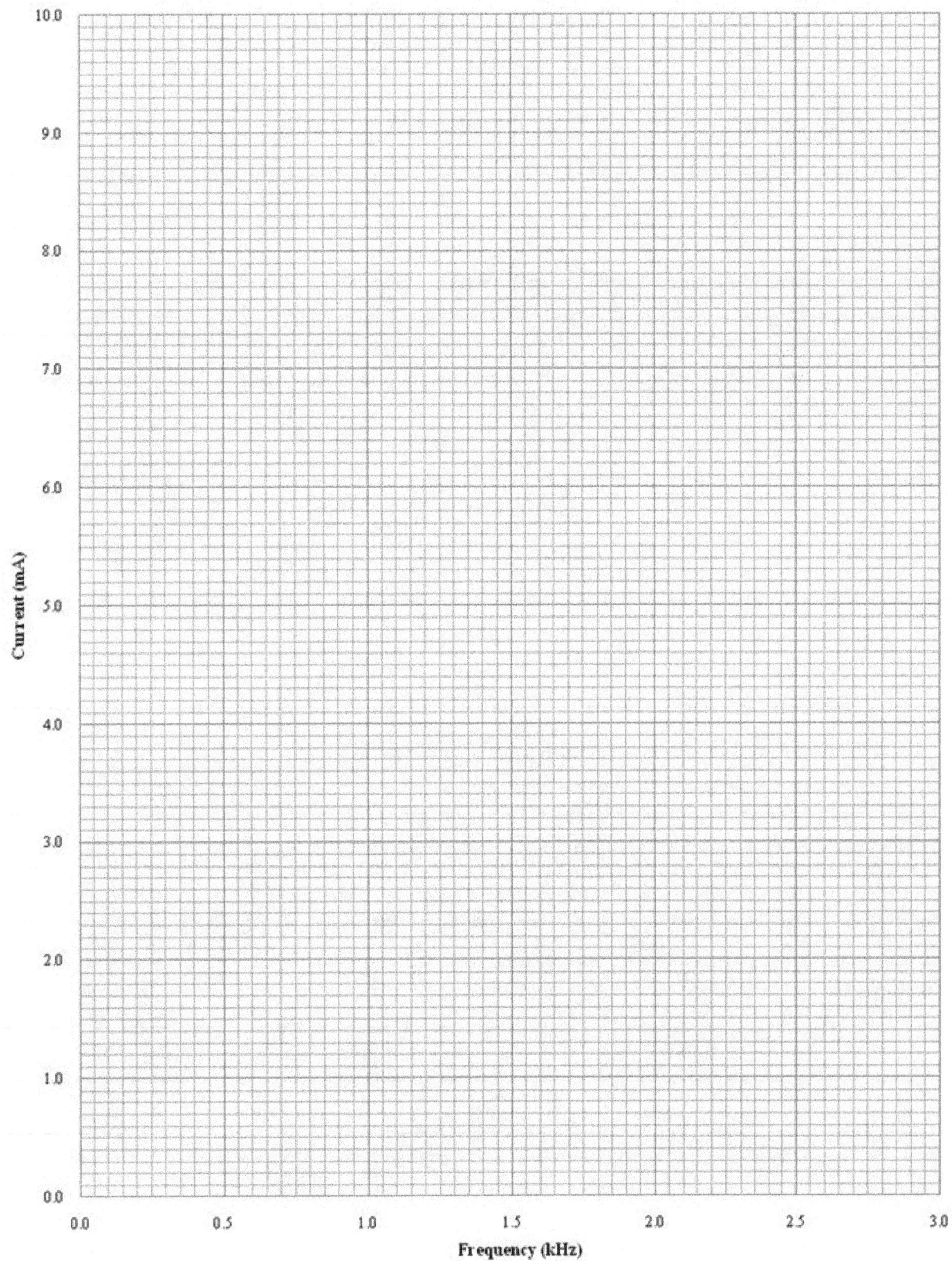
10.0
9.0
8.0
7.0
6.0
5.0
4.0
3.0
2.0
1.0
0.0
Current (mA)
0.0
0.5
1.0
1.5
2.0
2.5
3.0
Frequency (kHz)

QUESTIONS

1. Comment on the shape of your three resonance curves as related to the values of L and C.

2. What other factor (besides L and C) affects the bandwidth of a tuned circuit? Explain your answer.

3. Explain how bandwidth affects the selectivity of a tuned circuit.

4. Compare your measured values of bandwidth with your theoretical values. Discuss possible errors.

5. From your graph, determine the current flowing at resonance. How does this compare to the theoretical value? Explain any errors.

6. A series *RLC* circuit has a bandwidth of 15 kHz and a Q of 50 at resonance. A 10-mV source delivers a current of 100-μA at resonance. Determine the components and cut-off frequencies.

EXPERIMENT

27

PASSIVE FILTERS

OBJECTIVES

1 To investigate the properties of a low-pass filter.

2 To investigate the properties of a high-pass filter.

3 To investigate the properties of a band-pass filter.

4 To plot bode diagrams for filter circuits.

DISCUSSION

Passive filters are circuits, comprising combinations of inductors, capacitors, and resistors, which control the frequency response of an AC circuit. Since the reactance of an inductor and a capacitor is frequency dependant, circuits containing these components, along with resistors can be used to remove unwanted frequencies from AC signals.

Low-Pass Filters

A low-pass filter will reject frequencies above a certain value (*critical frequency*) and allow lower frequencies to pass through. The frequency response curve of a typical low-pass filter is shown below.

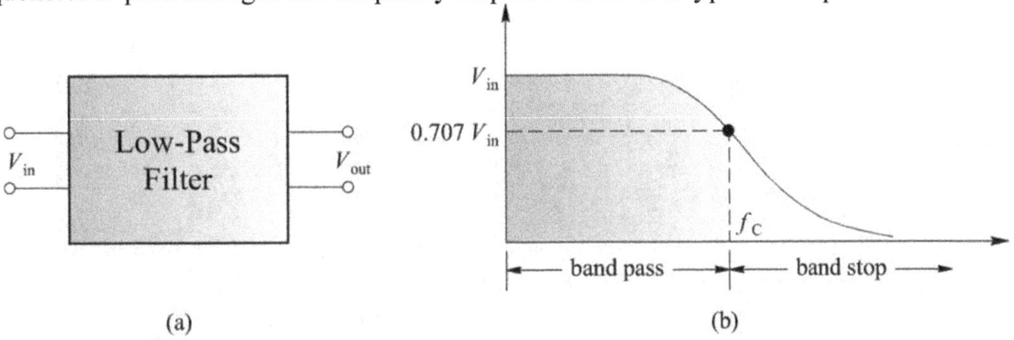

(a) (b)

The critical frequency (f_C) is the frequency at which $V_{out} = 0.707\ V_{in}$. Frequencies below this are assumed passed and frequencies above are rejected.

The output of a filter is normally expressed as a ratio of output to input voltage called the *voltage gain* (A_V). When expressed in terms of a logarithm it is called a *decibel* (dB).

$$dB = 20 \log \left(\frac{V_{out}}{V_{in}} \right)$$

A simple *RC* series circuit will attenuate higher frequencies and allow lower frequencies to pass through.

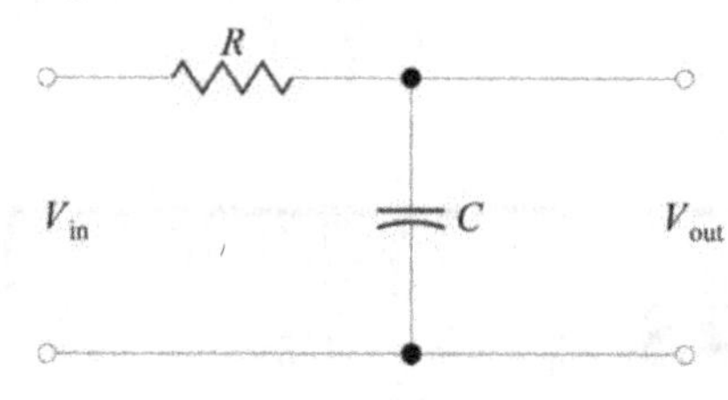

At low frequencies, the reactance of the capacitor is large and a greater proportion of the input voltage appears at the output. As frequency increases, X_C decreases and smaller voltages appear across the output. The frequency response curve of this practical filter is shown in the following diagram.

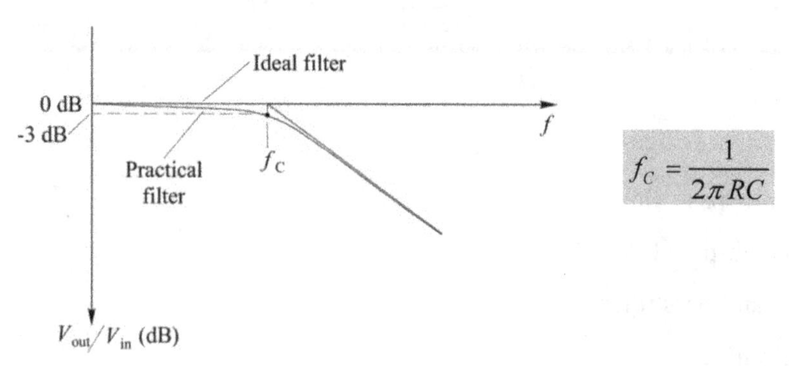

$$f_C = \frac{1}{2\pi RC}$$

The vertical scale is shown in terms of gain in dB and indicates the roll-off in output as the frequency is increased. At the critical frequency, the gain is −3 dB which represents a 50% reduction in power.

Bode Plot

As frequency increases above the critical frequency the output voltage continues to decrease. For each tenfold increase in frequency above f_C there is a 20 dB attenuation of the voltage. A tenfold change in frequency is called a *decade* and the roll-off in output is a constant −20 dB/decade for RC and RL filters. Because frequency response plots are normally done over three decades of frequency changes, it is common to plot the frequency on a logarithmic scale along the horizontal axis and the output on a linear scale on the vertical axis. This is called a *semilog* scale and the resulting plot is called a *Bode* plot.

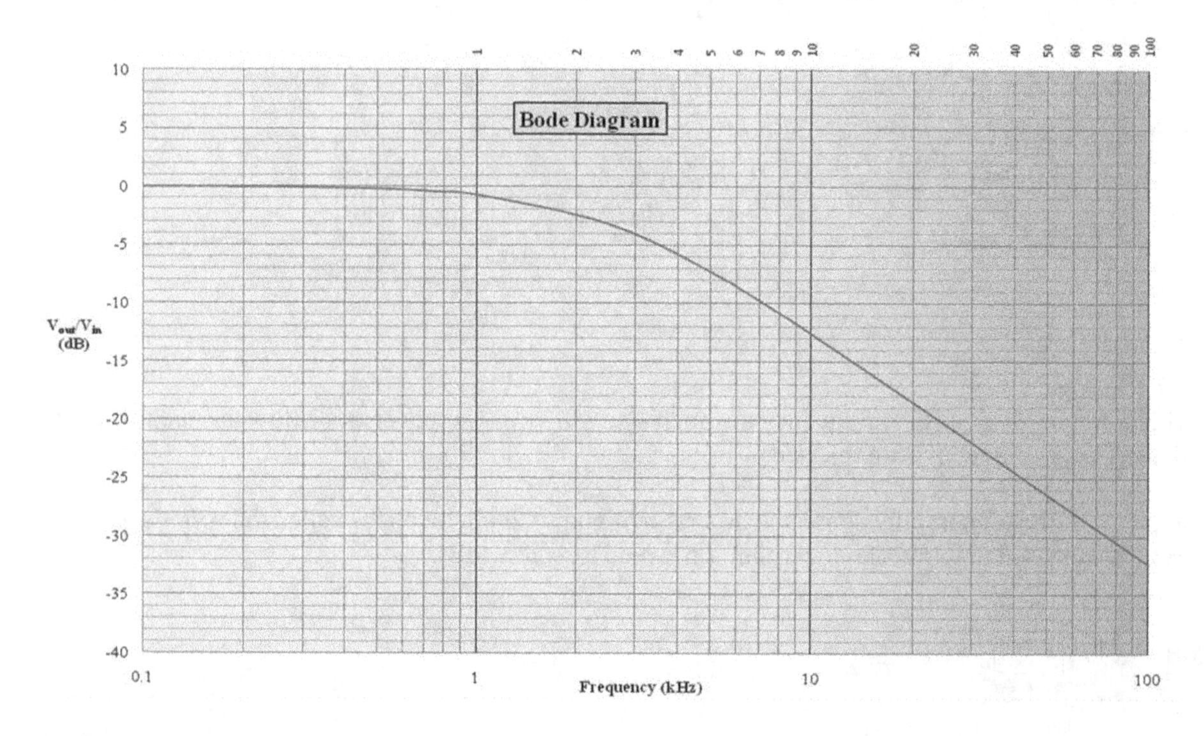

RL Low-Pass Filter

A similar graph will result from a series *RL* circuit. In this case the output is taken across the resistor. At low frequencies, X_L is small and most of the voltage appears across the resistor. As frequency increases, X_L increases and output voltage decreases.

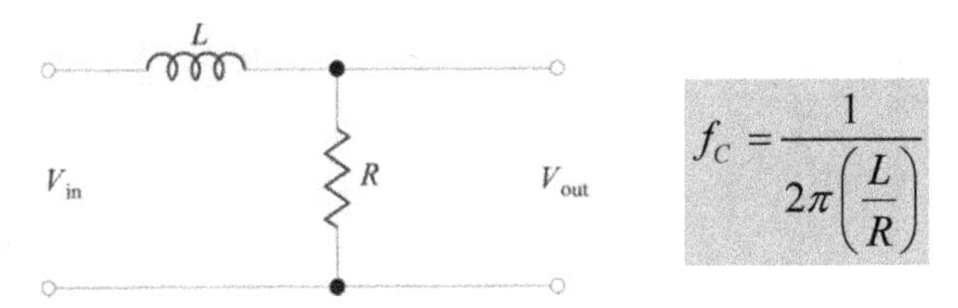

$$f_C = \frac{1}{2\pi\left(\frac{L}{R}\right)}$$

High-Pass Filters

A high-pass filter will reject frequencies below the critical frequency and allow higher frequencies to pass through. The frequency response curve of a typical high-pass filter is shown below.

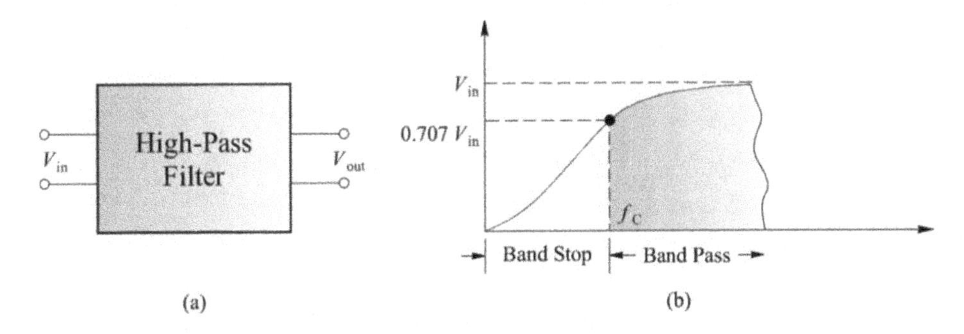

(a) (b)

Simple series *RL* and *RC* circuits can again be used as high-pass filters.

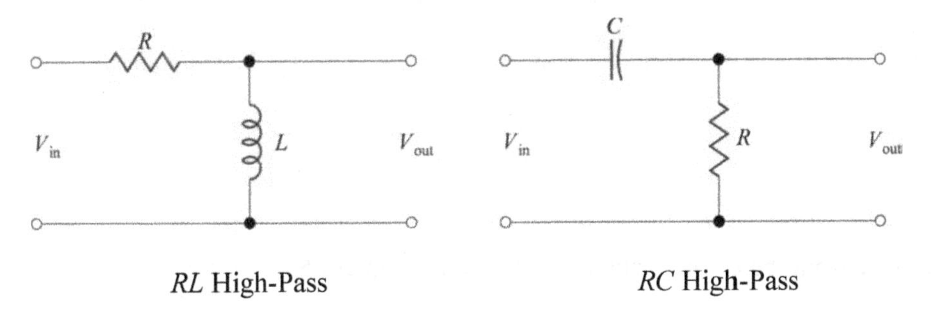

RL High-Pass *RC* High-Pass

In an *RL* circuit, X_L is small at low frequencies and therefore V_{out} is small. As frequency increases, X_L increases and the output increases. In the *RC* circuit, X_C is large at low frequencies and again V_{out} is small. X_C decreases with frequency increase and more voltage appears at the output for the higher frequencies. The formulas for f_C are the same as for the low-pass filter.

Band-Pass Filters

A band-pass filter allows a range of frequencies to pass and rejects frequencies above and below this band.

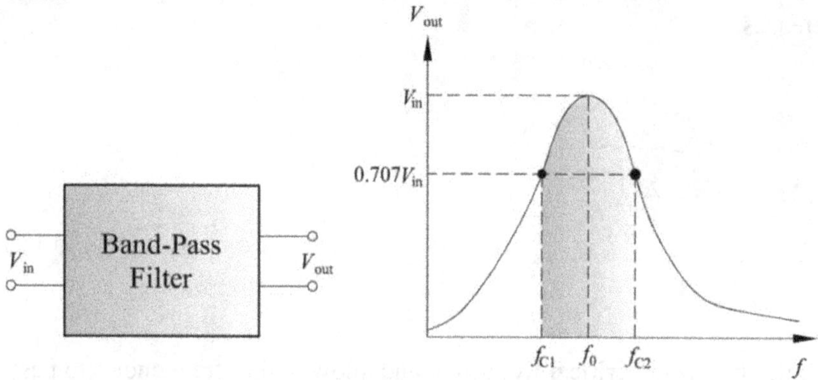

The simplest band-pass filter is the series resonant filter.

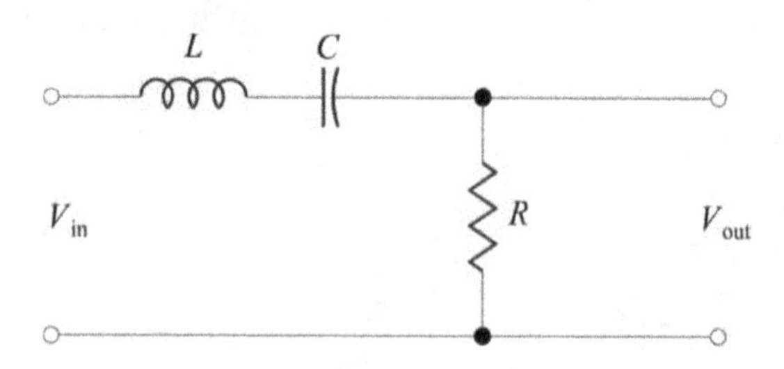

At low frequencies, X_L is small and X_C is large and only a small voltage appears across the resistor. As frequency increases, X_L increases and X_C decreases which reduces the net reactance and increases the output voltage across the resistor. At resonance $X_L = X_C$ giving zero net reactance and all the input voltage appears across the resistor. Beyond the resonance point, $X_L > X_C$ and V_{out} begins to decrease again. The *center frequency*, f_0, is actually the resonant frequency for the circuit and is given by the resonant frequency formula.

$$f_0 = \frac{1}{2\pi\sqrt{LC}}$$

The pass bandwidth is determined from the Q factor and f_0 as shown in Chapter 25.

$$BW = \frac{f_0}{Q}$$

EQUIPMENT

Power Supplies:
- Function generator

Instruments:
- DMM

Resistors:
- One each of 1.0 kΩ, 3.3-kΩ

Capacitors:
- One each of 0.047-μF and 0.0047-μF

Inductors:
- One each of 25-mH and 50-mH

Other:
- Protoboard
- Hook-up Wire

PROCEDURE

PART A: *RC* Low-Pass Filter

1. Connect the circuit shown in Figure 27-1.

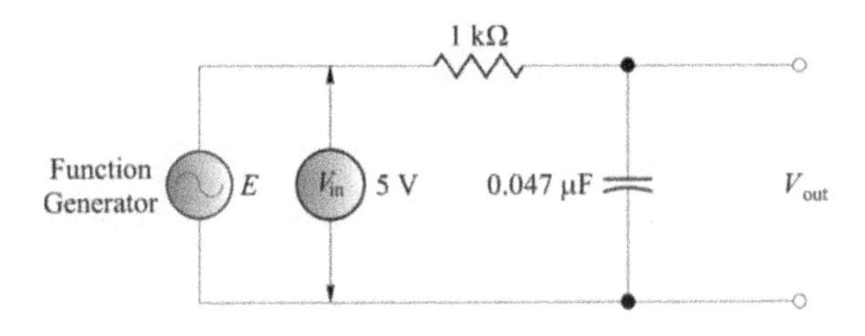

Figure 27-1

2. Switch **on** the function generator. Adjust the frequency to 100 Hz and the output to 5 V RMS. **This voltage must be maintained for each step of this experiment.**

3. Measure the voltage across the capacitor and record the value in Table 27-1.

4. Adjust the frequency to each of the values shown in Table 27-1 and repeat step 3. **Make sure that the source voltage is maintained at 5 V_{RMS} for each step.**

5. Calculate the ratio of $\frac{V_{out}}{V_{in}}$ for each frequency and express the gain in decibels. Record the values in Table 27-1.

6. Plot the Bode diagram on the semilog graph paper provided.

PART B: *RL* High-Pass Filter

7. Connect the circuit shown in Figure 27-2. Plug the function generator into the isolation transformer.

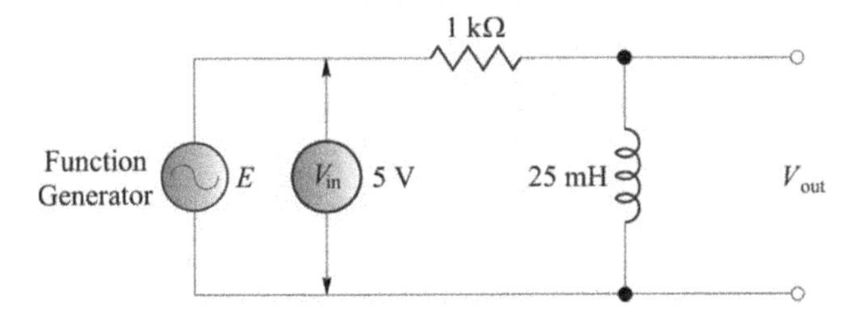

Figure 27-2

8. Switch **on** the function generator. Adjust the frequency to 100 Hz and the output to 5 V RMS. **This voltage must be maintained for each step of this experiment.**

9. Measure the voltage across the inductor and record the value in Table 27-2.

10. Adjust the frequency to each of the values shown in Table 27-2 and repeat step 9. **Make sure that the source voltage is maintained at 5 V_{RMS} for each step.**

11. Calculate the ratio of $\frac{V_{out}}{V_{in}}$ for each frequency and express the gain in decibels. Record the values in Table 27-2.

12. Plot the Bode diagram on the semi-log graph paper provided.

PART C: *RLC* Band-Pass Filter

13. Connect the circuit shown in Figure 27-3. Plug the function generator into the isolation transformer.

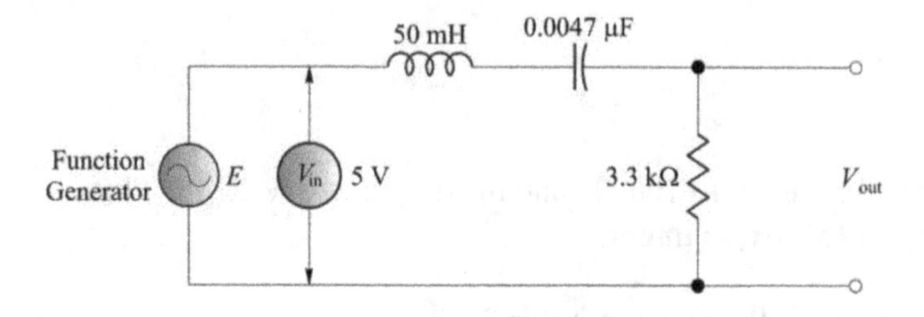

Figure 27-3

14. Switch **on** the function generator. Adjust the frequency to 100 Hz and the output to 5 V RMS. **This voltage must be maintained for each step of this experiment.**

15. Measure the voltage across the resistor and record the value in Table 27-3.

16. Adjust the frequency to each of the values shown in Table 27-3 and repeat step 15. **Make sure that the source voltage is maintained at 5 V_{RMS} for each step.**

17. Vary the frequency until the voltage across the resistor is a maximum. This is the center frequency (f_0). Record this value and the corresponding maximum voltage in table 27-3.

18. Calculate the ratio of $\frac{V_{out}}{V_{in}}$ (A_v) for each frequency. Record the values in Table 27-3.

19. Plot the frequency response diagram on the graph paper provided.

RESULTS

Table 27-1 *RC* Low-Pass Filter

f (kHz)	V_{In} (V)	V_{Out} (V)	$\Delta_V = \frac{V_{out}}{V_{in}}$	$\Delta_{V_{dB}} = 20log\Delta_V$, dB
0.1	5			
0.5	5			
1	5			
2	5			
5	5			
10	5			
15	5			
20	5			
50	5			
75	5			
100	5			

Table 27-2 *RL* High-Pass Filter

f (kHz)	V_{In} (V)	V_{Out} (V)	$\Delta_V = \frac{V_{out}}{V_{in}}$	$\Delta_{V_{dB}} = 20log\Delta_V$, dB
0.1	5			
0.5	5			
1	5			
2	5			
5	5			
10	5			
15	5			
20	5			
50	5			
75	5			
100	5			

Table 27-3 *RLC* Band-Pass Filter

f (kHz)	V_{In} (V)	V_{Out} (V)	$\Delta_V = \frac{V_{out}}{V_{in}}$	$\Delta_{V_{dB}} = 20log\Delta_V$, dB
0.1	5			
0.5	5			
1	5			
2	5			
5	5			
10	5			
12	5			
15	5			
20	5			
50	5			
75	5			
100	5			
f_0	5			

Frequency Response of the *RC* Low-Pass Filter

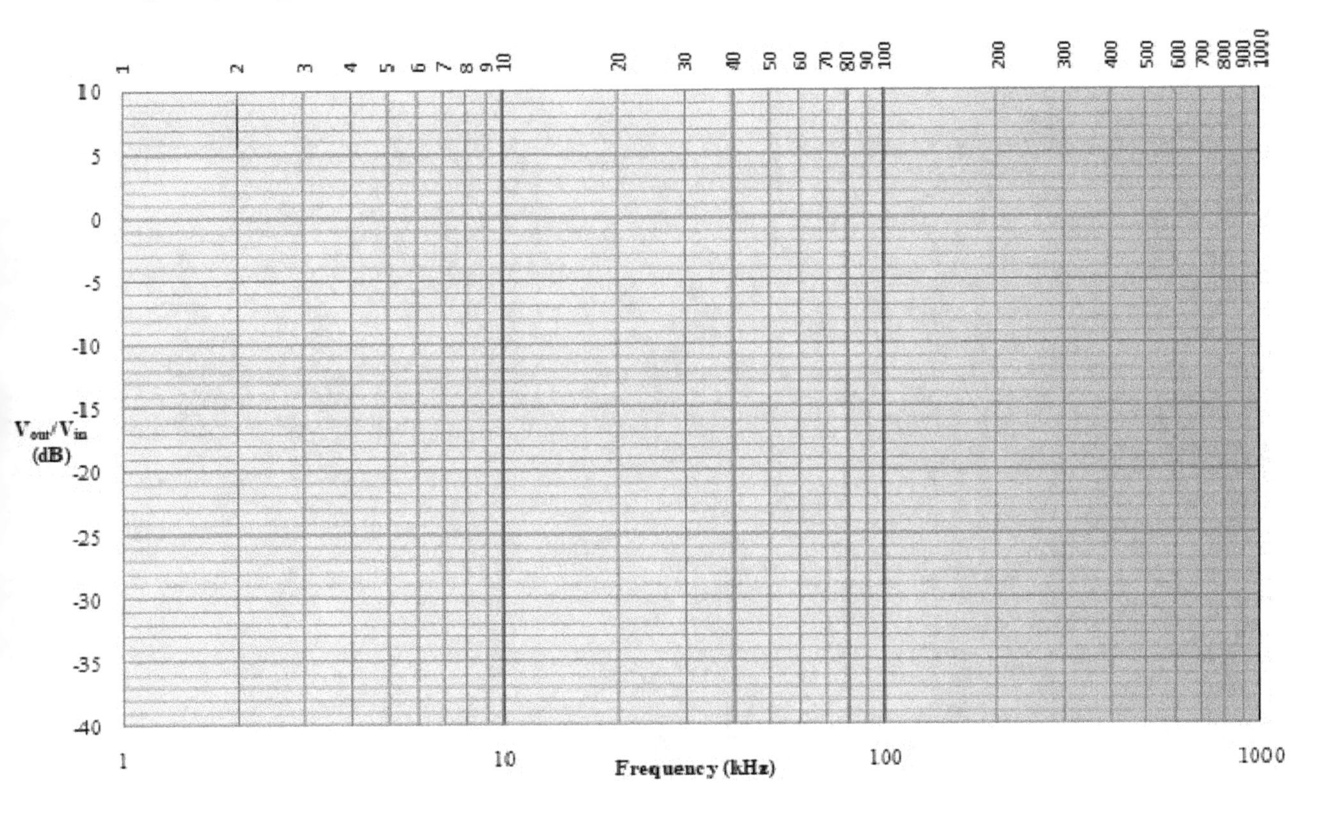

Frequency Response of the *RL* High-Pass Filter

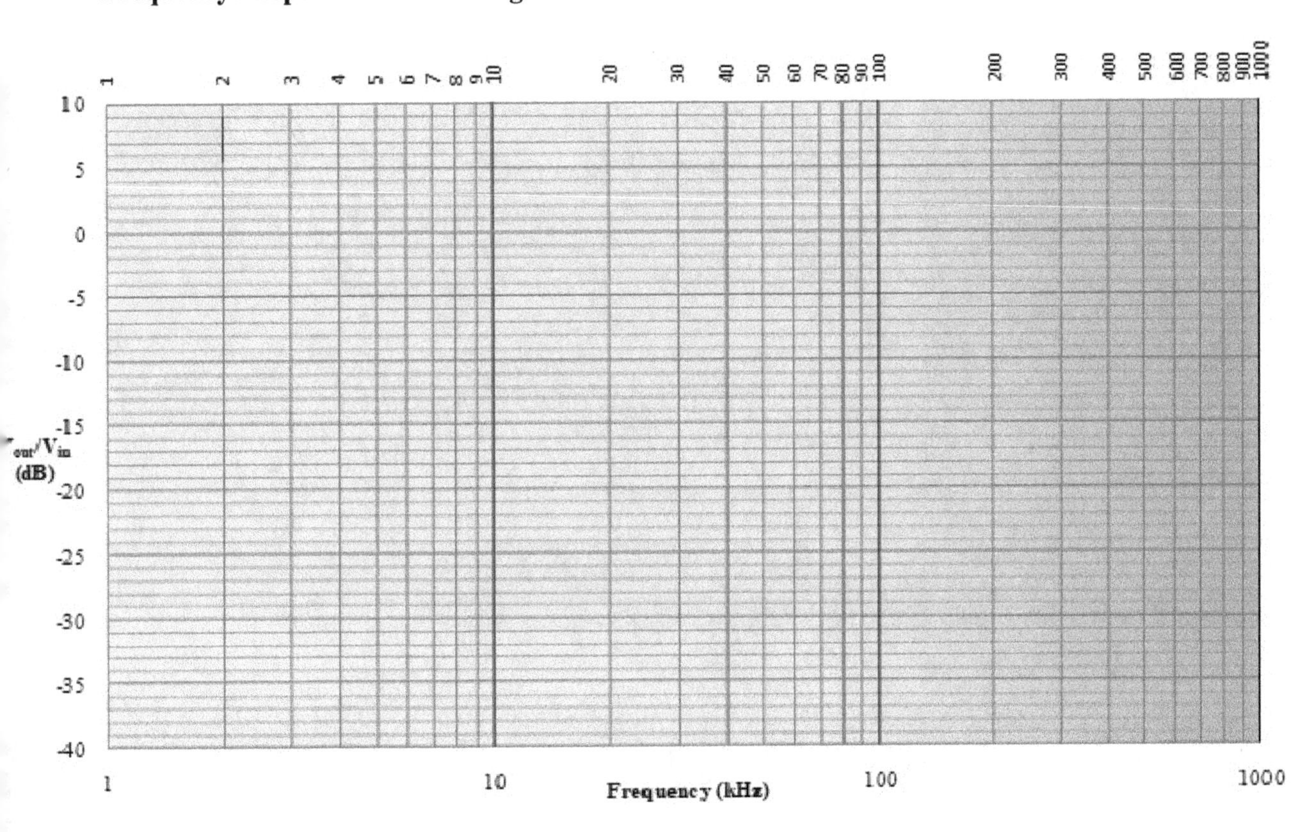

Frequency Response of the *RLC* Band-Pass Filter

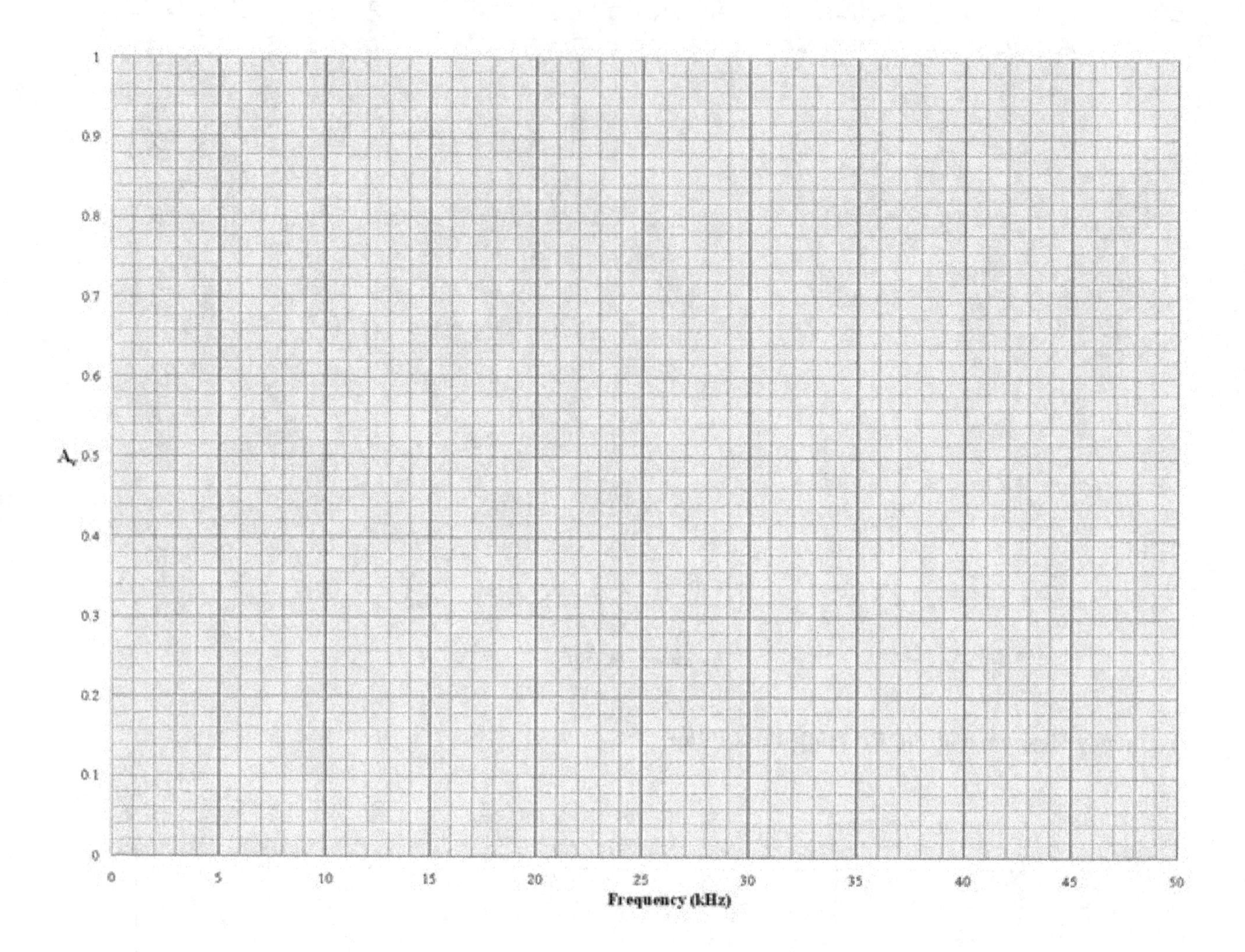

QUESTIONS

1. From your graph for Part A, determine the critical frequency and compare it with the theoretical value. Explain any discrepancies.

2. Determine the slope of the linear part of the Bode diagram for Part A in terms of dB/decade. How does this compare with the theoretical value?

3. From your graph for Part B, determine the critical frequency and compare it with the theoretical value.

4. Determine the slope of the linear part of the Bode diagram for Part B in terms of dB/decade. How does this compare with the theoretical value?

5. From your graph for Part C, determine the center frequency and compare it with the theoretical value. Explain any discrepancies.

6. Show the pass-band on your Bode diagram for Part C. Determine the bandwidth and compare it with the theoretical value.

7. Which of the circuits investigated in this lab would be best suited for a sub-woofer filter in a speaker system? Explain your answer.

8. A 20-pF capacitor is to be used in a band-pass filter with a center frequency of 125 kHz. Determine the inductor value to produce this filter. If a bandwidth of 10 kHz is required, what value of resistance must be added in series?

9. Use the Bode Plotter in Multisim to generate the frequency response of the *RC* Low-Pass Filter, the RL High-Pass Filter and the *RLC* Band-Pass Filter. Compare the frequency response of the simulation with the frequency response obtained with your three graphs generated in the lab.